BACTERIOLOGY RESEARCH DEVELOPMENTS

BIOLOGICAL AND CHEMICAL STUDIES TO SUPPORT THE USE OF LACTOBACILLI AS A STRATEGY FOR CONTROL OF BIOFILM-PRODUCING BACTERIA

BACTERIOLOGY RESEARCH DEVELOPMENTS

Additional books in this series can be found on Nova's website under the Series tab.

Additional E-books in this series can be found on Nova's website under the E-book tab.

BACTERIOLOGY RESEARCH DEVELOPMENTS

BIOLOGICAL AND CHEMICAL STUDIES TO SUPPORT THE USE OF LACTOBACILLI AS A STRATEGY FOR CONTROL OF BIOFILM-PRODUCING BACTERIA

J.C. VALDEZ
N.M GOBBATO
M.M. RACHID
A.N. RAMOS
A. BOSCH
M.C. PERAL
M.A. HUAMAN MARTINEZ
AND
M.L. GONZÁLEZ

Nova Science Publishers, Inc.
New York

Copyright © 2011 by Nova Science Publishers, Inc.

All rights reserved. No part of this book may be reproduced, stored in a retrieval system or transmitted in any form or by any means: electronic, electrostatic, magnetic, tape, mechanical photocopying, recording or otherwise without the written permission of the Publisher.

For permission to use material from this book please contact us:
Telephone 631-231-7269; Fax 631-231-8175
Web Site: http://www.novapublishers.com

NOTICE TO THE READER

The Publisher has taken reasonable care in the preparation of this book, but makes no expressed or implied warranty of any kind and assumes no responsibility for any errors or omissions. No liability is assumed for incidental or consequential damages in connection with or arising out of information contained in this book. The Publisher shall not be liable for any special, consequential, or exemplary damages resulting, in whole or in part, from the readers' use of, or reliance upon, this material. Any parts of this book based on government reports are so indicated and copyright is claimed for those parts to the extent applicable to compilations of such works.

Independent verification should be sought for any data, advice or recommendations contained in this book. In addition, no responsibility is assumed by the publisher for any injury and/or damage to persons or property arising from any methods, products, instructions, ideas or otherwise contained in this publication.

This publication is designed to provide accurate and authoritative information with regard to the subject matter covered herein. It is sold with the clear understanding that the Publisher is not engaged in rendering legal or any other professional services. If legal or any other expert assistance is required, the services of a competent person should be sought. FROM A DECLARATION OF PARTICIPANTS JOINTLY ADOPTED BY A COMMITTEE OF THE AMERICAN BAR ASSOCIATION AND A COMMITTEE OF PUBLISHERS.

Additional color graphics may be available in the e-book version of this book.

LIBRARY OF CONGRESS CATALOGING-IN-PUBLICATION DATA

Biological and chemical studies to support the use of lactobacilli as a
strategy for control of biofilm-producing bacteria / J.C. Valdez ... [et al.].
   p. cm.
  Includes index.
  ISBN 978-1-61728-859-3 (softcover)
  1. Lactobacillus. 2. Probiotics. 3. Biofilms. I. Valdez, J. C.
  QR82.L3B56 2010
  579.3'7--dc22
                      2010033118

Published by Nova Science Publishers, Inc. † New York

CONTENTS

Preface		vii
Chapter 1	Introduction	1
Chapter 2	Basic Studies	13
Chapter 3	In Vitro Assays	15
Chapter 4	Ex Vivo Assays	43
Chapter 5	Clinical Studies	59
Chapter 6	Conclusions	73
Acknowledgments		77
References		79
Index		93

PREFACE

Chronic wounds, by definition, are those that remain in a chronic inflammatory state and therefore fail to follow normal patterns of the healing process. The chronic wounds present a challenge to physicians and patients alike because they are very difficult to heal, inflict a huge cost to society and impair the quality of life for millions of people. There are many factors, that contribute to the development of chronic wounds. One of the most clinically significant impediments to wound healing is infections.

The growth of large aggregates of cells on a surface encased within a matrix of extracellular polymers produced by the sessile bacteria is known as a biofilm. In man, one of the surfaces that are available for attachment is ulcer beds. Bacteria biofilm formation in wounds is the best unifying explanation for the failure of chronic wounds to heal. It has been estimated that biofilms are associated with 65 percent of nosocomial infections. Biofilm bacteria are protected from host defenses, antibiotics, and antiseptics.

Using good bacteria to obstruct bad ones—a strategy known as bacterial interference—is one application of so-called probiotics, a field with growing medical promise. Beyond using probiotics to replenish missing bacteria on digestive tract and vagina, we and other authors are working to interfere to pathogenic bacteria in infected wounds. Furthermore, it has been demonstrated in vitro that cells and/or the metabolic by-products of lactobacilli have antagonistic effects on pathogens; these results have also been found in vivo during trials with urinary and genital infections in humans and mice and in wounds infected with S. aureus.

We demonstrated that *L. plantarum* interferes with the pathogenic capacity of *P. aeruginosa* (quorum sensing, biofilm, virulence factors, and growth).

Topical treatment with *L. plantarum* cultures is currently being carried out by our medical team with infected burns and chronic venous ulcers in humans with encouraging results. The cells from ulcer beds collected after treatment with *L. plantarum* showed a decrease in the percentage of polymorphonuclear, apoptotic and necrotic cells and a modulation of IL-8 production. The mode of action of *L. plantarum* on infection and wound healing seems to be related to acidity, presence of lactic acid, interference with *P. aeruginosa* quorum-sensing signals, perhaps inter-species communication through type 2 autoinducer and modification of tissue repair process.

Our findings will be useful for the formulation of effective and inexpensive products to resolve infected chronic wounds with biofilm-producing bacteria.

Chapter 1

INTRODUCTION

Cutaneous wound healing is a complex process involving blood clotting, inflammation, new tissue formation, and finally tissue remodeling.

This process of wound healing is extremely complex, involving hundreds of growth factors, dozens of integrins, scores of enzymes, and over ten different cell types.

Polimorphonuclear leukocytes especially neutrophils (PMN) are an essential component of the acute inflammatory response and the resolution of microbial infection. Infections, tissue injury and surgical trauma activate the release of inflammatory prostaglandins and chemoattractants (chemokines such as interleukin-8 and lipids as leukotriene B4) recruit PMN. Acute inflammation is characterized by rapid influx of blood neutrophils on encountering bacteria; neutrophils engulf these microbes into a phagosome, which fuses with intracellular granules to form a phagolysosome. In the phagolysosome the bacteria are killed after exposure to enzymes, antimicrobial peptides and reactive oxygenspecies (ROS).The arsenal of cytotoxic agents has been traditionally divided into either oxygen-dependent or oxygen-independent mechanisms. Both these systems collaborate in killing microbes.

After entering tissue granulocytes, platelets-leukocytes interaction promotes the switch of arachidonic acid-derived prostaglandins (E2and D2) and leukotriene B4 to lipoxines A4 and B4 as well as newly lipids generated from omega-3 poyunsaturated fatty acid (resolvins and protectins), which serves as a stop signal by blocking the further recruitment of PMN from the post capilarvenules and indicates the acute inflammation resolution. Neutrophil recruitment ceases and death by apoptosis is engaged. The uptake of apoptoitc cells by macrophages stimulated them to release anti-inflammatory and reparative citokines such as transforming grow factor-B1 required for wound (Sherjan and Savill 2005).

Granulation tissue is formed as fibroblasts produce extracellular matrix and endothelial cells form blood vessels. Extensive growth of epithelial cells, fibroblast deposition of collagen fibers in random patterns, and continued growth of blood vessels. Finally, collagen fibers become more organized, blood vessels are restored to normal, the scab is shed, and the epidermis is restored to normal thickness. The end result of uncomplicated healing is a fine scar with little fibrosis, and a return to near normal tissue architecture and organ function. If a wound does not heal in an orderly or timely sequence, or if the healing process does not result in structural integrity, then the wound is considered chronic.

Chronic wounds, by definition, are wounds that remain in a chronic inflammatory state and therefore fail to follow normal patterns of the healing process. The chronic wound presents a challenge to physicians, caregivers, and patients alike because are very difficult to heal, inflict a huge cost to society and impair the quality of life for millions of people (Lazarus et al 1994).

There are many factors that contribute to the development of chronic wounds. Similarly, the failure of these wounds to heal is also multifactorial. Some of the most commonly encountered and clinically significant impediments to wound healing include wound infection, hypoxia, presence of debris and necrotic tissue, use of anti-inflammatory medications, a diet deficient in vitamins or minerals, or general nutritional deficiencies, and metabolic disorders, such as diabetes mellitus. Chronic wounds are rarely if ever sterile and achieving wound sterility is often an unrealistic and nonessential goal of wound care. More important than the physical presence of bacteria is the number of organisms present per gram of tissue in a wound and the type of organisms identified. The critical number of bacteria able to cause a wound infection has been shown to be 10^5 bacteria per gram of tissue for most bacteria. If 10^5 organisms are present per gram of tissue, the chances of successful wound closure are low (on the order of 19%). If, on the other hand, 10^5 organisms per gram of tissue are present, then the likelihood of successful wound closure is approximately 94% (Stadelmann et al 1998). However, the precise interaction between microbes in the wounds and impaired healing is unknown. The microflora of ulcers is usually polymicrobial and recent studies using molecular techniques have emphasized the complex ecology of these wounds and it was observed that 86% of ulcers with no clinical signs of infection contained more than one bacterial species. Today, the interaction between ulcer and bacteria can be stratified into four levels: contamination, colonization, critical colonization, and infection. Whilst, contamination and colonization by microbes are not believed to inhibit healing, the line between colonization and infection can be difficult to define. Moreover, the underlying pathogenesis of chronic wounds may result in wounds of different aetiologies being differently affected by bacteria and a range of clinical criteria have been used to define infection in chronic wounds (Howell-Jones et al 2005).

In regard to these concepts, we should take account that there is increasing evidence to believe that biofilm formation in wounds is the best unifying explanation for the failure of chronic wounds to heal.

Biofilms are dense aggregations of microbial cells attached to a surface. These surface-attached communities are known to have a significant impact on human health when they form on medical and surgical implants.

The mechanism of the biofilm formation begins when free-living bacteria (planktonic phase) detect the availability of nutrients or other benefits, then set about synthesizing a substrate biofilm and virulence factors and forming real communities; 85% of the biofilm's extracellular matrix and only 15% are cells with a particular metabolism that prevents the antibiotics act on their targets. Numerous laboratories have made the observation that after a majority of a biofilm population has been killed by an antimicrobial, a very small percentage of the population remains viable despite prolonged exposure to the antibiotic or increased dosage. These cells are called "persisters" and confer no heritable resistance to progeny once the selective pressure. Biofilm resistance is based on persister survival (perhaps a small bacterial number). An initial treatment with antibiotic kills planktonic cells and the majority of biofilm cells. The immune system kills planktonic persisters, but the biofilm persister cells are protected from host defenses by the exopolysaccharide matrix. After the antibiotic concentration drops, persisters resurrect the biofilm and the infection relapses (Lewis 2001)

Anecdotal clinical evidence indicates improved healing when chronic wounds are treated with the assumption that biofilm is the cause of the failure to heal. This treatment can comprise a number of approaches: use of antibacterial biofilm agents, debridement, anti-biofilm dressings, phages, biocides and advanced technologies.

The syntheses of the biofilm and virulence factors are regulated by chemical signals secreted by the bacteria. When bacteria reach a threshold number, there is also a threshold concentration of signals that trigger the expression of genes involved in biofilm synthesis and virulence factors. Thus the activity of these signals depends on cell density (quorum) and, thus, is called quorum sensing signals. Both Gram-negative and Gram-positive pathogens are known to use autoinducer molecules to coordinate expression of genes crucial for virulence and survival. Delaying virulence factor production until a certain population density is reached might allow for a host to be overwhelmed before its innate immune response is fully activated. In addition, this process might enable bacteria to function in concert with the characteristics of a multicellular organism by acquiring an organization such as a biofilm which can aid to pathogenesis.

The autoinducer AI-1 and the systems that are involved include: a) the system LuxI/luxR for Gram (-) bacteria where quorum sensing signals are acyl-homoserine lactones (AHLs) produced by an enzyme system LuxI and detected by a LuxR system which are proteins that, when dimerized by the autoinducer (AHLs), bind to the gene promoter DNA and activate transcription of genes coding for the synthesis of the biofilm and virulencia factors; b) the system of oligopeptides for Gram (+) where bacteria synthesize and transport oligopeptides as autoinducer, called AIPs. These consist of 5 to 27 amino acids. The detection circuit is made using a two-component signal transduction, leading to phosphorylation of a regulatory protein which binds to a promoter DNA and regulates gene transcription in biofilm and virulence factors.

There are other signs that allow bacteria to recognize each different species to determine their behavior. These autoinducer called AI-2 (chemically are furanone), involved in their synthesis the Lux-S enzyme so that any bacteria that encoded in their genes (http // www.ncbi.nlm.nih.gov / BLAST), can have an interspecies communication to have synergistic or antagonistic behaviors. For example, enzymes such as AHL-lactonase produced by the genus Bacillus, calls AiiA, hydrolyze lactone ring joints. Strains of *S. aureus* interfere with quorum sensing signal activity of other strains of S. aureus. The algae Delisea produces Furanones that have antimicrobial properties because displace AHL to its receptor LuxR. Strategies that interfere with communication in bacteria are being explored in the biotechnology industry with the aim of developing novel antimicrobials (Federle and Blassler 2003).

In the infection caused by microorganisms producer of biofilm Polymorphonuclear leukocytes (PMNs) are the first line of defense and they are very efficient to eliminate these pathogens in planktonic fase by a process known as phagocytosis. During phagocytosis, PMNs produce reactive oxygen species (ROS) and release cytotoxic granule components into phagocytic vacuoles to kill ingested microbes. PMNs accumulate rapidly at sites of infection; therefore, the regulation of PMN turnover following phagocytosis is important for preventing damage to healthy tissues that would otherwise occur in the event of necrotic cell lysis. During the course of microbial infections, signal transduction events that facilitate host cell responses are relayed through PMN surface receptors, including those specific for complement receptors (CRs) and Antibodies (FcRs), IL-8–10, IL-13, IL-15, and IL-18, and multiple receptors for bacteria and microbial products, such as the Toll-like receptors (Muzio et al 2000).

When in infection bacteria grow in biofilms polymorphonuclear neutrophils (PMNs) are attracted to biofilms and may penetrate these sessile communities. PMNs can enter into the water channels of a biofilm. However, although the membranes of these phagocytes remain intact, they seem to be "paralyzed" in that they have not internalized any bacteria; so, PMNs invade biofilms but are virtually inactive in killing sessile cells and resolving biofilm infections. Also, individual infected with these bacteria shown height title of specific antibodies, but their antimicrobial activity is no apparent (Costerton et al 1999).

The innate immune system evolved several strategies of self/non self discrimination that is based on the detection of conserved microbial molecular patterns that are essential products for their physiology. These conserved structures are referred to as pathogen-associated molecular patterns (PAMPs) and are recognized by receptors of the innate immune system called pattern recognition receptors (PRRs).

Bacteria or bacterial products (PAMPs) elicit a strong inflammatory host response. This response is mediated principally by the innate immunity cells. A class of PPR expressed on the cell surface, the Toll-like receptor (TLR) family, is central regulators of the innate immune response: they respond to conserved molecular patterns shared by bacteria, fungi and viruses. TLR2 either as a homodimer or heterodimer with TLR1 or TLR6 interacts with lipoteichoic acids, peptidoglycans and other compounds. TLR4 is activated by endotoxin and also recognizes pneumolysin. Short segments derived from bacterial DNA stimulate cells of the innate immune system via TLR9. Stimulation of TLRs activates macrophages in the reticuloendothelial system and circulating leukocytes via a common pathway including the adaptor molecule MyD88, and via MyD88-independent mechanisms including the adaptor molecules TIRAP, TRIF and TRAM. This eventually leads to the release of proinflammatory cytokines, chemokines, matrix metalloproteinases and nitric oxide (Nau and Eiffert 2005). Host-bacteria interactions involved multiple molecular recognition events that are required for productive infection by pathogen or for resistance against infection by the host. It is important that PPRs do not distinguish between pathogenic or comensal bacteria and all of them produce PAMPs, but only pathogens evolved the means to gain access to the compartments within the host mediated virulence factors. These factors are not necessary for bacterial growth, but function to subvert the host defense system and cause disease. Deletion of virulence factors renders pathogenic microbes avirulent. Virulence factors induce host tissular injury with cell death by necrosis or apoptosis and release endogenous products, (the non foreing alarm signal in the danger model) such as mammalian DNA, RNA, heat shock proteins (Hsps), breadown products of hyaluron heparin sulfate, fibronectin and fibrinogen. The cells of innate immune system such as dendritic cells, macrophage-monocites and polymorphonuclears can response to both injury/pathogen-related signal and normal physiological signals involved with apoptosis induced by pathogens and of course to PAMPs. The recognition of these products lead to the activation of transcription nuclear factors such as NFkB, production of mediators of inflammation and the activation of antigen-presenting cells (APCs) that initiate the adaptive (specific) immune response. (Rusland and Janeway 2002; Kanzler et al 2007). Commensal bacteria interact with cells, and of course, with immune cells but differently to pathogenic bacteria. The paradigmatic example is the human intestinal epithelia. This is continuously exposed to the luminal microbiota. In healthy humans, this exposure to non-self stimuli activates balanced and interconnected innate and adaptive immune responses leading to tolerance toward nonpathogenic bacteria. Tolerance is a general term describing the state by which the immune system is rendered non reactive toward self or non-self antigens (Clavel and Haller 2007). We are an environment to an uncountable number of symbiotic, commensal and pathogenic organisms, each of which has had evolutionary time to learn how to use and misuse our immune system. Each organ is a complex combination of tissues, delicately balanced to perform a particular function: a function that can easily be compromised by the powerful effector mechanisms wielded by the immune system. Thus, tissues use all sorts of mechanisms to keep the cells and molecules of the immune system out until they need them and to control them when they arrive. Whereas pathogens induce mainly persistent inflammatory responses, commensal bacteria induce transient, non-inflammatory responses (Matzinger 2007).

Using harmless bacteria to displace pathogenic organisms is an alternative and promising way of combating infections. Using good bacteria to obstruct bad ones—a strategy known as bacterial interference—is one application of so-called probiotics. Folk remedies since the beginning of recorded history have ascribed antidiarrheal effects to dairy foods, such as cheeses and yogurt.

It turns out that their active component is live bacteria. The effect of probiotics on digestive tract health is a field with growing medical promise. Other scientists recognize potential applications to the vagina, the respiratory tract, and other parts of the body where bacterial ecosystems also exist. For example, studies by Sharon L. Hillier of Magee-Women's Hospital in Pittsburgh suggest that probiotic lactobacilli that researchers isolated from the vaginas of healthy women can help combat an infection called bacterial vaginosis, which is caused by anaerobic microbes that trigger a shift in vaginal flora. This infection affects about 1 in 10 U.S. women of reproductive age (Harder 2002). Beyond using probiotics to replenish missing bacteria, clinicians are working to prevent pathogenic invasions of normally bacteria-free tissues that are, for one reason or another, at risk of infection. The strategy in these cases is to install harmless bacteria at the site before nasty ones invade.

Given the many bacteria that employ quorum sensing in the control of virulence and biofilm formation, both autoinducer AI-1 (intraspecies) and AI-2 (Interspecies) quorum sensing constitutes a novel target for directed drug design (Federle and Blassler 2003). This approach to treatment is not new. Bacterial interference was once widely studied, and attempts to influence colonization of pathogenic bacteria with "harmless" bacteria were carried out some decades ago. It is know as Bacteriotherapy. In human health bacteriotherapy was probably forgotten because of the continuous development of new, more potent antibacterial agents and because of fears about possible side effects. Avirulent bacterial strains can, in principle, also cause infections (Tagg and Dierksen 2003). Roos et al showed how commensal haemolytic streptococci were used to replace the normal nasopharyngeal flora in children with recurrent otitis media. The results were astonishing. After treatment, recurrences of otitis media fell to half of those in the control group; at three months 42% of children given streptococci in nasal spray were healthy compared with 22% of the controls; and the need for new courses of antibacterial treatment decreased. Roos et al have also successfully used haemolytic streptococci in preventing recurrent streptococcal tonsillitis (Roos et al 2001). Reid et al. experimented with using probiotic lactobacilli to prevent Staphylococcus aureus from colonizing animals' wound sites, where it can cause life-threatening infections. They simulated an infection during surgical operation by implanting small pieces of silicone simultaneously added *S. aureus* beneath the skin of laboratory rats. Rats that received large doses of one live *Lactobacillus fermentum* strain at the wound site during surgery remained free of the pathogen (Gan et al 2002).

Another target for probiotic therapy is the bladder. Hull of Baylor College of Medicine in Houston and Darouiche of New York Medical College-Westchester Medical Center in New York began a decade ago to use probiotics to prevent recurrent bladder infections. Unlike the gut or vagina, a healthy bladder doesn't contain microorganisms, which makes introducing bacteria to the organ a counterintuitive idea that's unpopular with some clinicians. However, the bladder is not sterile in this patient group (people with spinal-cord injuries that render them unable to urinate without a catheter). The catheter tube makes occasional infections almost inevitable. To protect a patient against such infections, Hull sends a small amount of solution containing an apparently harmless strain of *Escherichia coli* through the catheter and into the bladder. This inoculation won't fight an infection already in progress, but it does make it more difficult for new harmful bacteria to take hold (Darouiche et al 2001; Darouiche and Hull 2000; Prasad et al 2009). Direct instillation of lactobacilli into the bladder of individuals with spinal cord injury and their impact on immune system is under evaluation (Anukam et al 2009).

It is now accepted that all these chronic infections are produce by pathogenic bacteria because their ability to make biofilm, an observation that is not weighted properly.

The other cause would be to consider is the interaction of components of harmless bacteria and pathogenic bacteria on the defense mechanisms of host.

Based on our research we hypothesized that one cause of the benefit that exercise harmless bacteria (*Lactobacillus plantarum*) and a chronic wound infected by pathogens (*Pseudomonas aeruginosa*) that related to the interference and / or inhibition of their quorum sensing molecules, virulence factors and biofilm production of the pathogen and the impact this has on host defense mechanisms. (Valdez et al 2005). Increasing evidence, including human studies, is also supporting the immunomodulatory role attributed to given lactic acid bacterial strains. The immunomodulating effect of lactic acid bacteria (LAB) on the gut associated lymphoid tissue (GALT) (van Baarlena et al 2009) and the ability to signal the host's own defenses, such as mucins in the gut defensins and antimicrobial immune factors in the gut and vagina, is likely to be very important (Reid et al 2006).

The direct interaction of nonpathogenic bacteria with human PBMC was based on the assumption that bacteria and immunocompetent cells may physically interact in definite mucosal environments. The investigation of interactions of lactobaccilli with leukocytes in vitro was made to evaluate the immunomodulatory capacity of intestinal bacteria immune cells and to investigate if a distinct pattern of immunomodulation can be established for different components of the microflora. The ease of obtaining blood immune cells compared with information obtained from the mucosal sites immune cells particularly in humans generated a great deal of research in which researchers studied the direct interactions of lactobacilli and blood immune cells both in vitro and in vivo.

In vitro, se determine que several species of lactobacilli have been shown to induce cytokines, such as tumor necrosis factor alpha (TNF-α) and interleukin-12 (IL-12) in human peripheral blood mononuclear cells (PBMC) (Hessle et al 1999). IL-12, like gamma interferon (IFN-γ), is an important cytokine implicated in innate defense mechanisms in response to bacteria (Haller et al 2000).

It has been reported that live probiotic lactobacilli are more potent inducers of cytokine production in mammalian leucocytes than killed forms. Live *L. casei* Shirota was capable of inducing J774A.1 cells to secrete marginally higher levels of IL-12 and TNF-α than those observed using a comparable number of heat-killed bacilli; however, more noteworthy was the fact that even live *L. casei* Shirota did not induce IL-10 production in J774A.1 cells, indicating that – overall – the innate pattern recognition of this probiotic strain does not influence IL-10 production (Cross et al 2004).

Hessle et al. has reported that in vitro assay gram-positive and gram-negative bacteria induce different patterns of immunoregulatory cytokines in human monocytes: gram-positive bacteria induced on average 18 times more IL-12 than IL-10, while gram-negative bacteria on average induced 7 times more IL-10 than IL-12. Interleukin-10 and IL-12 are two cytokines secreted by monocytes/macrophages in response to bacterial products which have largely opposite effects on the immune system. IL-12 activates cytotoxicity and gamma interferon (IFN-γ) secretion by T cells and NK cells, whereas IL-10 inhibits these functions. The ratio IL-12/IL-10 for *L.plantarum* was 14 and for *P. aeruginosa* was 0.06. (Hessle et al 2000). Monocytes and PBMC stimulated with Gram-negative bacterial species induced much more PGE2 than did Gram-positive bacteria. PGE2 production induced by *P.aeruginosa* was four times larger than that induced by *L.plantarum*. (Hessle et al 2003).

As mentioned above, individual Toll-like receptors activate common and unique transcription factors through different signaling pathways to drive specific biological responses against microorganisms. TLR4, which was demonstrated to be the long-sought receptor for bacterial lipopolysaccharide (LPS), the outer membrane component of Gram-negative bacteria and TLR2 recognizes peptidoglycan, in addition to the lipoproteins and lipopeptides of

Gram-positive bacteria and mycoplasma lipopeptide. TLR2 appears to collaborate with its relatives TLR1 and TLR6 to discriminate between the molecular structures of diacyl and triacyl lipopeptides, respectively. Furthermore, TLR2, in concert with non-TLR receptors, even further diversifies its recognition potential (Kawai and Akira 2005).

To current knowledge, *L. plantarum* is perceived via TLR 2–4, CD14 antigen, and nucleotide-binding oligomerization domain-containing 2 (NOD2) (Karlsson et al 2004; Hasegawa et al 2006).

Intriguingly, although *L. plantarum* may induce innate or adaptive mouse immune responses, with production of pro inflammatory cytokines as TNF-α and IL-12 by PMNC (Karlsson et al 2002), it effect in vivo was shown to be anti-inflammatory; has been reported that whereas pathogens induce mainly persistent inflammatory responses, (Mayer-Scholl et al 2004) commensal bacteria induce non-inflammatory responses. Although, differential immune responses toward living and dead *L. plantarum* are described showed that s.c. mouse injections with preparations of viable *L. plantarum* induced mild inflammatory responses, whereas dead bacteria induced antibody formation by plasma cells and infiltration of polymorphonuclear cells (Bloksma et al 1979). The pathogens have mechanisms to damage the cells to subvert the defense responses. The molecules released from damaged cells can act as danger signals. It was suggested that there is a category of damage- associated molecular patterns (DAMPs) that encompasses both PAMPs and alarm signals. Commensal bacteria like *L. plantarum* lacking virulence factors does not damage cells and its effect on the immune system is reduced to its structural characteristics without release of danger signals. (Matzinger 2007)

Also there is in vivo study as well as in animal models and in human assays where the immunomodulatory effect of lactobacillus given by oral via was reported.

We studied in vivo the effect of yogurt on the inhibition of colon tumors induced by 1,2-dimethylhydrazine in BALB/c mice, and we established that yogurt induces a great reduction in the inflammatory immune response and inhibits tumor growth. We suggest that one of the mechanisms by which yogurt exerts antitumor activity is through its immunomodulator activity, by reducing the inflammatory immune response, which was markedly increased when the carcinogen was administered (Perdigon et al 1998).

Although, we determined in vitro assays that *Lactobacillus delbrueckii* and *Streptococcus thermophilus* were able to inhibit the apoptosis of macrophages induced by *S. typhimutium*, *L. delbrueckii* ssp. *bulgaricus* being the more effective one. The results suggest that these microorganisms could play a role in apoptotic mechanisms, since the inhibition of it would avoid pathogen dissemination. The preventive effect of LAB against *S. typhimutium* infection could be mediated by apoptosis inhibition (Valdéz et al 2001).

In human healthy was investigated the immunomodulatory effects in intestinal lamina propia after ingestion of of living and heat-killed commensa *L. plantarum* bacteria. Biopsies were taken from the intestinal duodenal mucosa. The authors shown that *L. plantarum* modulates in vivo transcriptional profiles of the proximal small intestinal mucosa involved in innate and adaptive immune responses.

After consumption of *L. plantarum*, genes that prevent overt adaptive immune responses were induced, whereas genes involved in establishing and amplifying inflammatory immune responses were not expressed or not modulated. Antigenicity of the *L. plantarum* wall TA depends on its association with the protein fraction and only consumption of dead *L. plantarum* leads to increased expression of TNF-α and of genes involved in T cell activation, antigen processing and presentation. In addition, only consumption of dead *L. plantarum* induced expression of genes involved in attraction and maturation of T cells such as the genes encoding the interleukins 2, 7, and 8. However, the consumption of *L. plantarum*, induced genes associated with anti-inflammatory activities, potential NF-kB inhibitors A20, IkB, BCL3, and SOCS3 (suppressor of cytokine signaling). No infiltration of immune cells was observed in human biopsy sections. These striking differences between live and dead *L. plantarum* comsuption it depends of differences in modulation of NFkB pathways. Association of different subunit of NFkB determines activation of genes to different cytokines (van Baarlena et al 2009).

This new point of view supports bacteriotherapy scientifically using harmless bacteria to antagonize with infectious pathogens that its ability to form biofilm are difficult to eradicate in the wounds.

However, the empirical use of *L. plantarum* on patients infected wounds not resolved by conventional treatments, with remarkable results began in 1988 in Plastic Surgery and Burns Unit of Hospital Health Center "Zenon Santillan of the City of Tucuman, Argentina, relied on the ability of lactobacilli to inhibit in vitro growth of pathogenic bacteria, especially *P. aeruginosa* that commonly cause infections in our hospitals.

The hospital is one of the main state-supported reference health care centers in the northwestern area of Argentina.

Since in 2000 we started doing basic and clinical researches to determine the mechanisms by which *L. plantarum* could antagonize the pathogens that infect chronic wounds.

Chapter 2

BASIC STUDIES

Our basic investigation focuses on the effects of *L. plantarum* on *P. aeruginosa* in vitro, ex vivo and in vivo assays.

In vitro we studied the interference of *L. plantarum* on biofilm, quorum sensing signals, virulence factors and growth of *P. aeruginosa*.

Ex vivo we study the effect of supernatants of *L plantarum* and *P aeruginosa* in the intracellular pH, viability, apoptosis and necrosis in human PMN.

In vivo we study the effect of *L plantarum* has on infection by *P. aeruginosa* in a murine model of burns. We also studied the effect of subcutaneous inoculation of bacterial supernatants has on the skin of mice.

Pseudomonas aeruginosa is a Gram-negative opportunistic pathogen that infects primarily immunocompromised individuals, such as patients with cystic fibrosis, cancer or AIDS, or patients with indwelling medical devices or burns (Kievit and Iglewski 2000). Resistance to antimicrobial agents and numerous virulence factors means that *P. aeruginosa* causes many recalcitrant infections. *P. aeruginosa* virulence determinants include cell-associated (lipopolysaccharide endotoxin, flagellum, pili) and extracellular (alginate, exotoxin A, exoenzyme S, pyocyanin, elastase, etc.) factors. It has been demonstrated that expression of these virulence factors is regulated by a quorum-sensing system. This system has two components, las and rhl, and uses two autoinducers, N-(3-oxododecanoyl)-L-homoserine lactone (3O-C12-HSL) and N-butyryl- L-homoserine lactone (C4-HSL), respectively, to control expression of the virulence factors. The flagellum, pili and exopolysaccharide are important in pathogenicity because they allow the bacteria to form biofilms (Goldberg et al 2007). Host defenses such as phagocytes polymorphonuclear neutrophils (PMNs), the dominant cells of the inflammatory response, and macrophages can be intoxicated through the activity of the type III secretion system and other virulence factors of *P. aeruginosa* (Dechaeux et al 2000) and remained paralyzed by biofilm (Bjarnsholt et al 2005). Also, antibodies are inhibited to act against bacteria shielded into biofilm (Costerton 1999).

Chapter 3

IN VITRO ASSAYS

In these studies we found that both the whole cultures of *L. plantarum* as the culture supernatant had a strong antipathogenic activity on *P. aeruginosa*. This activity was represented by inhibition of four determining factors in the pathogenicity of *P. aeruginosa*: I) Inhibition of biofilm, II) Growth inhibition, III) Inhibition of quorum sensing and IV) Inhibition of virulence factors (elastase).

The *P. aeruginosa* strains used in this study were a standard clinical isolate. *L. plantarum* ATCC 10241 was grown in MRS broth (Oxoid, Basingstoke, UK) at 37°C.

The cell-free supernatant of *L. plantarum* ATCC 10,241 was obtained from an entire culture in MRS (24 hours at 37 °C) which was centrifuged and filtered with Millipore filter (0.22 µm). This supernatant has a pH= 4.95 ± 0.24 and was called acid supernatant of *L. plantarum* (SLp). SLp aliquots were neutralized with NaOH, thus obtaining the neutral supernatant (SLpN) (pH = 7.00).

BIOFILM OF *P. AERUGINOSA*

When *P. aeruginosa* adheres to a surface, the cells differentiate to form microcolonies covered by an extracellular matrix that eventually come to be a biofilm (Costerton et al 1994). Biofilms have a very complex structure, which can be identified channels through which exchange oxygen and other substrates with the aqueous phase. Cells within a biofilm are usually enmeshed in an extracellular matrix produced by the microorganism itself (Costerton et al 2003). This matrix is a complex mixture of exopolysaccharides (alginate) (Davies et al. 1993), proteins and DNA. DNA is derived from lysed cells (Webb et al 2003) and secreted through small DNA containing vesicles located in the outer membrane (Renelli et al 2004). The biofilm forming bacteria have a different metabolism of cells grown in liquid media (called planktonic), one of the more obvious differences is its decreased sensitivity to antibiotics and other toxic agents (Nickel et al 1985).

I) Anti-Biofilm Activity

The inhibitory activity of culture supernatants of *L. plantarum* on the biofilm of *P. aeruginosa* was clearly demonstrated through different methodologies. These methodologies threw each on its side similar percentages of inhibition and in some cases were compared with a typical inhibitor-disruptor of *P. aeruginosa* biofilm, the enzyme DNAase.

Inhibition of Biofilm Formation

In 96 Wells Microplates: (O'Toole and Kolter 1998)
It is well known that one of the principal components of the biofilm matrix of *P. aeruginosa* is DNA (from bacterial secretion and host). That's why a culture of *P. aeruginosa* on various beds of DNA were placed in these plates and incubated with the inhibitors: SLp, SLpN and human recombinant DNAase solution (Figure 1). As seen, the inhibitory capacity of SLp is higher than DNAase, a known inhibitor, in all beds used. However, the SLpN loses much of its inhibitory capacity, indicating a predominant role of the acidity of the supernatant in the inhibition mechanism.

Biofilm Disruption

Fluorescent Stains
In other studies with plates of 96 wells had also noticed a significant biofilm disrupting activity (data not shown). To confirm this, we placed separately in Petri dishes, beds of mentioned different DNA. Then were added a culture of *P. aeruginosa* to all of them and allowed to form biofilm. Finally were added separately the disruptors SLp or DNAase and incubated. Samples

were fixed with p-formaldehyde and then performed fluorescent staining techniques (DAPI for DNA) and direct immunofluorescence (anti-FITC-labeled Pseudomonas Rhodamine). In figure 2, it is clear that with DNAase, the most biofilm disruption occurs. SLp also shows some disruption capacity which is partly lost with the neutralization of the same. Hence, acidity plays a central role in *P. aeruginosa* biofilm disruption (Data not shown).

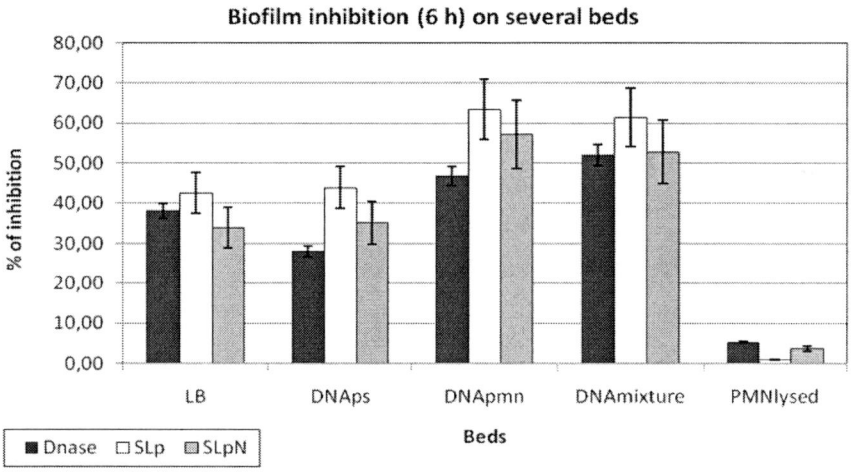

Figure 1. Inhibition of biofilm of *P. aeruginosa* grown on different beds of DNA (DNAps = genomic DNA of *P. aeruginosa*. DNApmn = polymorphonuclear genomic DNA. DNAmixture = mixture of equal parts of P. aeruginosa genomic DNA with genomic DNA from PMNs. lysed PMN = Polymorphonuclears lysates that have released their DNA but also their granules, as with those found in an ulcer *in vivo*.

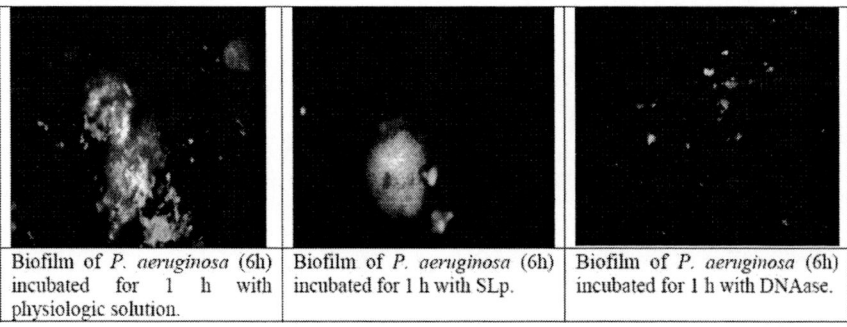

Figure 2. Mixed fluorescent stains that demonstrate the disruption capacity of SLp.

Kinetics of Biofilm Disruption with SLp in Continuous Culture

We performed a continuous culture system in which was used silicone tubes as substrate for biofilm formation. Whole system was filled with a culture of *P. aeruginosa* in LB, using a peristaltic pump. Once the system was filled, flow was stopped and incubated at 37 ° C for 4 h to promote adhesion. When finished that period of time, began pumping LB to allow biofilm formation. Samples were collected at 12 and 24 hours and then began pumping a mixture LB:SLp (4:1). Further samples were taken at 36 and 48 hours. Each sample taken was a piece of 8 cm in length. This piece was used to measure the amount of biomass (formed biofilm) on it by staining with crystal violet.

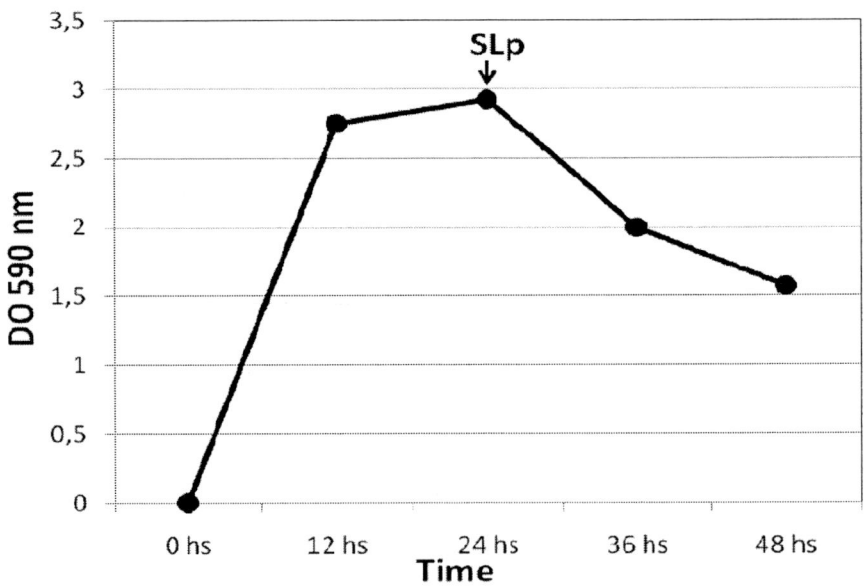

Figure 3. Curve of growth-biofilm disruption of *P. aeruginosa* in a continuous culture system. At 24 h began pumping the system with a mixture of LB:SLp (4:1).

As shown in **figure 3**, there is a steep slope of biofilm formation the first 12 hours to 24 hours decreases the slope because hose saturation. When finally starts pumping of mixture LB/SLp, shows a sharp decrease in biomass found in the silicone hose. This would indicate matrix destabilization and disruption of the biofilm formed during the first 24 hours.

Infrared spectroscopy is a nondestructive technique which allows obtaining information of the overall chemical composition of a sample (Nichols et al 1985; Naumann et al 1991; Schmitt and Flemming 1998). This technique has been used over the years in chemical analysis and in the characterization of biological samples. The ability of Fourier transform infrared (FT-IR) spectroscopy to discriminate structural and biochemical changes in the composition of microbial cells accompanying their surface-associated growth has made this technique particularly suitable for monitoring biofilm cultures (Nivens et al 1995).

The utilization of the mentioned technology, allowed us to visualize the chemical changes produced in the biofilm matrix of *P. aeruginosa* in presence of SLp. When we analyzed by FT-IR samples of biofilm formed on the hoses (before and after the addition of SLp), we found in all cases, peaks corresponding to those identified as characteristic of a *P. aeruginosa* mucoid strain. This strain produced biofilm that absorbed IR radiation near 1650 cm^{-1} (amide I), 1550 cm^{-1} (amide II), 1240 cm^{-1} (P=O stretching, C—O—C stretching, and/or vibration of amide III), 1100 to 1000 cm^{-1} (C—OH and P—O stretching) 1450 cm^{-1}, and 1400 cm^{-1}. Besides the biofilm of our strain, produced spectra with an increase in relative absorbance at 1060 cm^{-1} (C—OH stretching of alginate) and 1250 cm^{-1} (C—O stretching of *O*-acetil group in alginate) (Nivens et al 2001). We found no variations in the ratio of peaks with each other or for those matrix components (alginate, DNA). This would indicate that the biofilm disruption effect of the supernatant of *L. plantarum* is not based on attacking a particular component of the matrix, but the biofilm as a whole. It is important to note that the slope of disruption decrease after the first 24 hours of pumping LB:SLp. This could be because the bacteria would adapt to the hostile environment generated by the SLp.

Biofilm Inhibition Mechanism

Inhibition of Biofilm in Batch Culture Kinetics

We mentioned earlier that the acidity of the SLp was closely related to the inhibition mechanism of *P. aeruginosa* biofilm. One of the major components and most bring to the acidity of the SLp is lactic acid (LA). In order to elucidate the mechanism of inhibition we performed a biofilm growth kinetics in batch cultures on beads of polypropylene in the presence of different solutions that are aimed to assess: 1) Saline (NaCl 0,85% normal kinetics), 2) SLp; 3) SLpN 4) solution of lactic acid (LA) with the same concentration (130 mM) and pH as the SLp. To find out about which stages of biofilm formation acted these solutions, the kinetics were carried out: a) Adhesion: 1-4 hours; b) Latency: 5-10 hours c) Biofilm formation: 12-72 hours. All tubes were incubated at 37 ° C and taken in the different times mentioned. In each tube were performed the following tests:

Biomass

We evaluated biomass (biofilm) formed on the pearls in all cases by staining with crystal violet techniques. Figure 4 shows kinetic curves of biofilm formation up to 72 hours. In this figure we compare the normal kinetics (with saline) with those produced by the addition of LA, and SLpN SLp respectively.

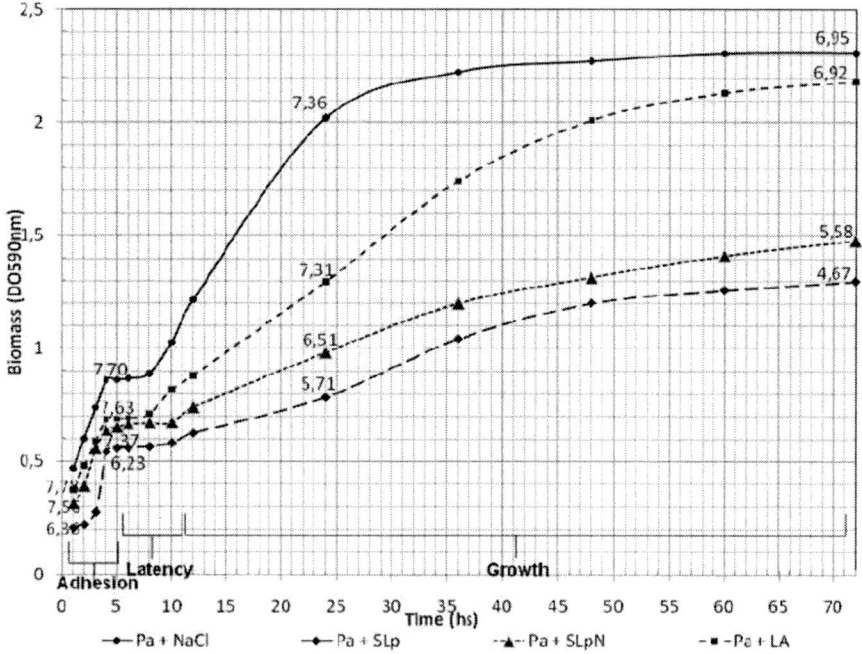

Figure 4. Kinetics of adhesion and biofilm formation on polypropylene beads in a batch culture in the presence of saline (normal kinetics), lactic acid (LA), PFS and SLpN. The numerical values observed on each curve correspond to pH.

All cases can be observed well-defined four periods. In the first periods, lasting approximately 5 hours, the cells are adhering to the pearls. During this time the lowest percentage of adhesion with respect to the normal kinetic was obtained by SLp, is followed by SLpN and finally LA which begins and ends with the same percentage of adherence (Figure 5).

In the second period, which we call latency (Figure 4), it did not show an increase of biomass over time. This is due to bacteria are phenotypically transforming for biofilm growth mode. This stage lasts 3 to 4 hours in normal kinetics and extends from 5 to 6 hours in the presence of SLp or SLpN. This gives us the pattern that in the supernatants of *L. plantarum* there would be some kind of molecules that interfere with normal phenotypic transformation.

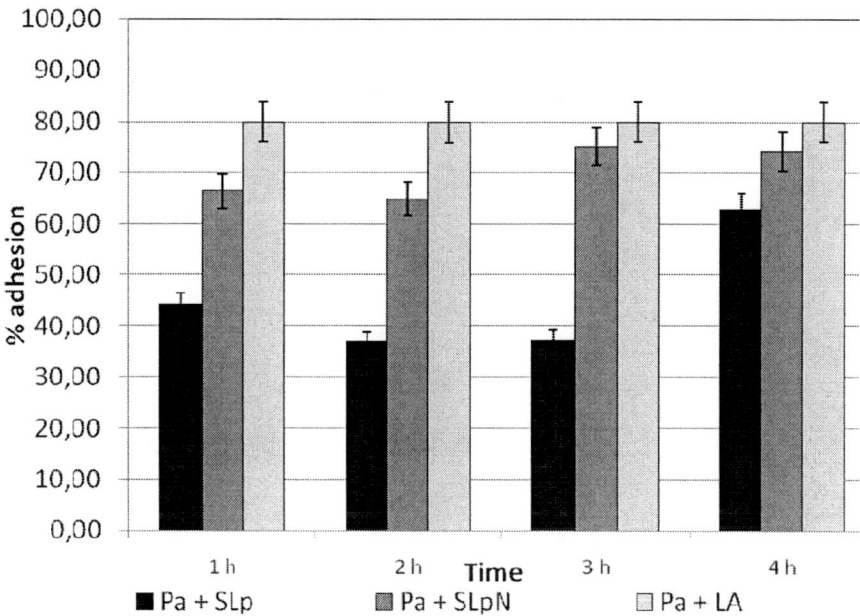

Figure 5. Percentage of adhesion represented as biomass. The percentages were calculated by taken as 100% the value for that time in the normal kinetics.

The third period is the formation of biofilm itself, through increased biomass formed on the pearls, reaching the saturation of the same in the fourth period. The steepest growth possesses normal kinetics and therefore reaches saturation more quickly (24 hours approximately). It follows the LA with a slope less steep and with a delay to reach saturation (72 h). The SLpN and SLp show a greatly reduced growth and fail to saturate the beads (Figure 4). This can be seen more clearly in figure 6 where we see that SLp leaves form 51% of normal biomass at 12 hours reaching 56% at 72 hours. With SLpN something similar happens with values ranging from 60% at 12 h to 63% at 72 hours. Apparently, this could be related to pH, because there is lower adhesion and biofilm formation at lower supernatant pH. In **figure 4** are pH values measured in time on each curve.

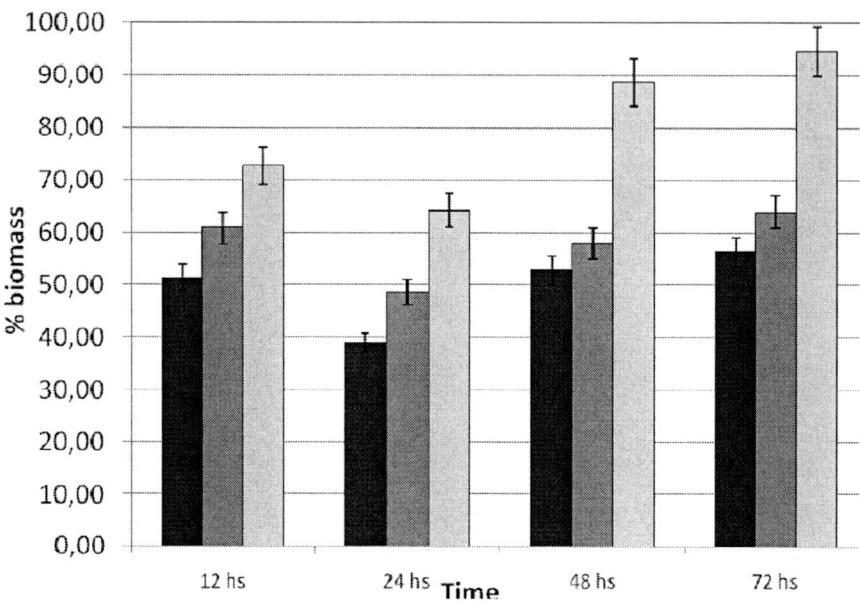

Figure 6. Percentage of biofilm formation represented as biomass. The percentages were calculated by taken as 100% the value for that time in the normal kinetics.

Chemical Composition of the Biofilm

To do this, we took off the biofilm formed on polypropylene beads by vigorous vortex in distilled water and then a homogeneous suspension was obtained. This suspension was placed in Zinc Selenide (SeZn) optical plates and moderate vacuum to obtain a transparent and homogeneous layer. All spectra were registered on a Perkin Elmer spectrometer model Spectrum One (USA). The readings were performed under dry air to reduce the water vapor contribution to measurements. When analyzing the FT-IR spectra at various times we note that the characteristic peaks of a biofilm mucoid strain of *P. aeruginosa* are present in all cases, and the relationship between peaks is the same. Making a computational analysis with the OPUS program version 4.0 we can say that bacterial growth (biomass) (represented by the peaks of amide I and II) and biofilm (represented by the peaks of alginate) are normal. This means that we are in presence of mature biofilm in each of the samples. However, when comparing time to time spectra obtained in the normal kinetics with those obtained in the presence of SLp, we see that the absorbance units are smaller. This would indicate that in the presence of SLp mature biofilm is formed but in smaller amounts. Put another way, the SLp only inhibits biofilm formation without affecting the composition of its matrix (Figure 7). The spectra obtained in the presence of SLpN and LA showed less inhibition (data not shown).

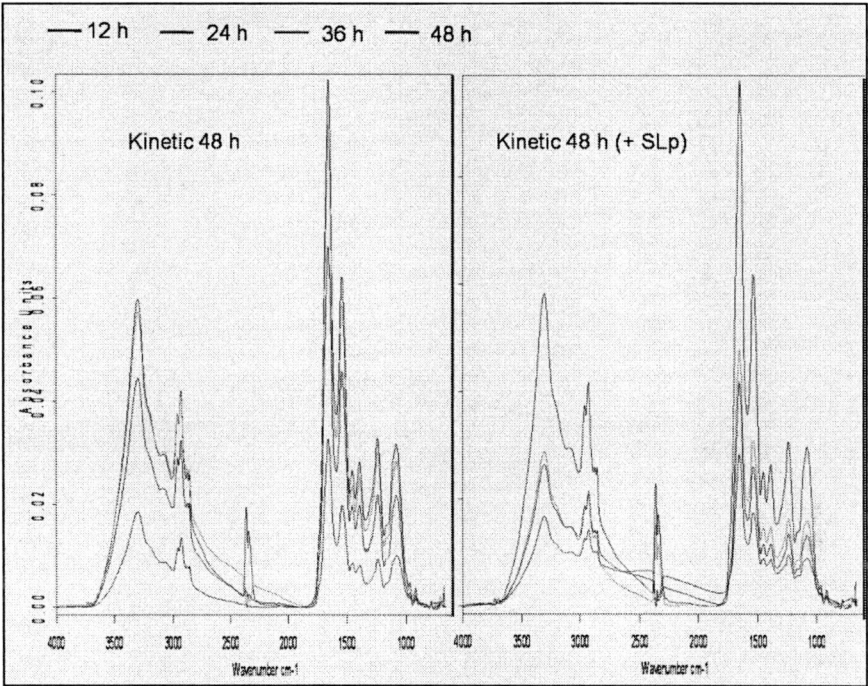

Figure 7. FT-IR spectra of biofilms obtained on polypropylene pearls at different times in normal kinetics (NaCl 0.85%) and in the presence of SLp.

Live-Dead Stains

By using a kit of bacterial viability (Live Syto9/ Dead IP BacLight Molecular Probes), we measured the percentage of live bacteria in the same homogeneous biofilm suspensions used for FT-IR. With the results we plot a curve of percentage of live bacteria vs. time (kinetics of bacterial killing). When we observe the death kinetics of bacteria in biofilm (Figure 8) we saw that as time goes on, who produces more mortality is LA reaching an 81.4% of live bacteria at 48 hours. The supernatants of *L. plantarum*, both the acid and the neutralized, produce death kinetics in biofilm and growth slopes directly proportional to pH at each time. It means that at lower pH greater percentage of dead bacteria in biofilm and lower growth slope. Interestingly, the curves for the SLpN are very similar to normal curves (NaCl 0.85%), which would indicate that the cause of death by bacterial supernatants is due exclusively to the acidity of them. But we can also say that the lactic acid itself (regardless of pH) also causes the death of *P. aeruginosa*.

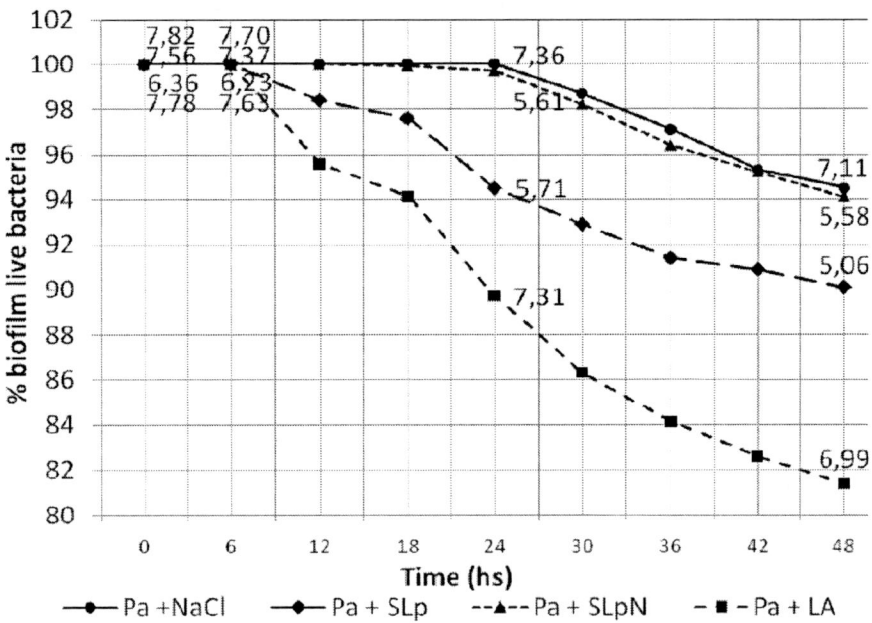

Figure 8. Dead Kinetic of biofilm bacteria measured by fluorescent staining with Syto 9-IP. The numerical values observed on each curve correspond to pH.

II) EFFECT ON GROWTH

CFU/mL and pH

In the supernatants of these systems aliquots were taken and measured the CFU/mL (successive dilutions and seeding in LB agar) and pH. This particular test shows the effect of the various solutions on the viability of planktonic cells and changes in pH on time. We see again in figure 9 that the effect of the supernatants on planktonic cells is directly proportional to its acidity, while the lactic acid solution produces a decrease in viability by a direct effect on cells.

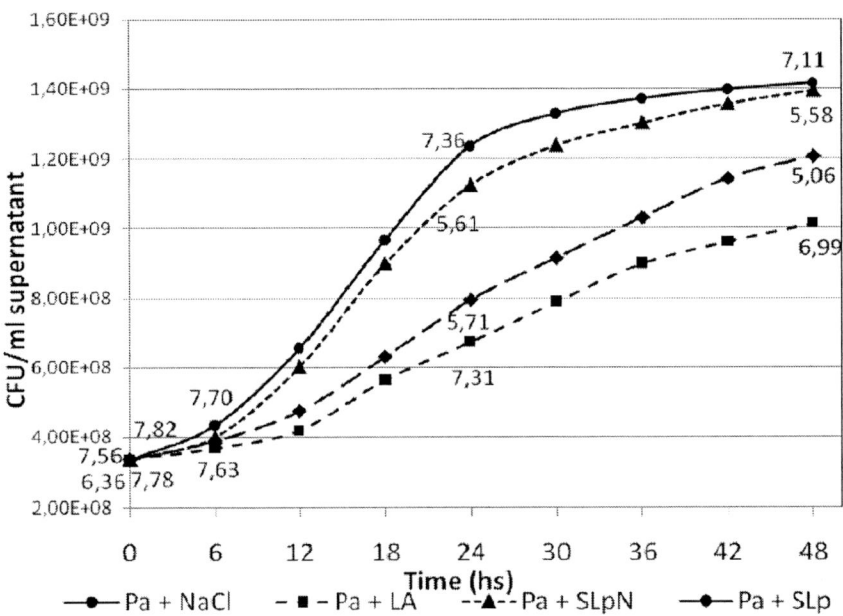

Figure 9. Growth bacterial curves represented by CFU / mL supernatant of batch cultures in the presence of various inhibitors. The numerical values observed on each curve correspond to pH.

In this work, through the inhibition assays in batch and continuous cultures, we could elucidate on what stage of biofilm formation of *P. aeruginosa* act *L. plantarum* supernatants. We note that affects both bacterial adhesion and biofilm growth itself, and even the phase of latency is delayed. However, microscopic observations show structures consistent with normal biofilm structure (data not shown). It can be observed bacterial adherence at first, then the appearance of microcolonies and subsequently mature biofilm formation.

One of our hypotheses about why all these inhibitory effects is the acidity of the SLp, as the neutralization makes it lose inhibition and disruption power, as can be seen in the trials with SLpN. However there is an extra inhibitory factor, still present in SLpN, as the same still retains inhibition power. It was noted also, that the SLp is capable to inhibit growth of *P. aeruginosa*. This could also be due to SLp pH, since the SLpN presents a curve of growth and death kinetics in biofilm almost normal.

A second hypothesis would explain the inhibition by the presence of lactic acid in the supernatants. Therefore measured the concentration in the SLp, and prepared a solution with the same concentration of lactic acid and the same pH as the SLp. The increased mortality of biofilm and planktonic cells, partially explain the mechanism used lactic acid to inhibit *P. aeruginosa*. But tests conducted with the above solution, is not reached inhibition values that SLp showing itself. Again this leads us to believe that there is an extra factor, present in the SLp, besides acidity and lactic acid that would produce much of the inhibition.

A third hypothesis we considered is the presence of specific enzymes against biofilm matrix components in SLp. It could be possible the presence of alginase, DNAase and even proteases to explain both the disruption and inhibition of biofilm formation. By looking at all the FT-IR spectra, we see that while the spectra obtained in the presence of SLp show a smaller amount of biofilm as a whole, the ratio of the peaks associated with DNA, alginate and proteins remains constant, indicating that there is not a direct attack on these components of the matrix. This is ratified in previous works (Valdez JC et al 2005; Ramos et al 2008), where we demonstrated the absence of DNAase and proteases against the biofilm matrix in the SLp.

The amount of the inhibiting factors found in the SLp: pH, lactic acid, mortality and inhibition of bacterial growth still fail to explain the significant degree of inhibition produced by supernatants of *L. plantarum* on the biofilm of *P. aeruginosa*. That's why we outlined a new hypothesis to explain this phenomenon.

III) INHIBITION OF QUORUM SENSING

Quorum Sensing System in *P. Aeruginosa*

Pseudomonas aeruginosa, like other bacteria, has multiple regulatory systems that allow modifying the expression of different genes in response to changes in the environment. Moreover, it is the bacteria in which it has found a higher proportion of genes that encode proteins that are potential transcriptional regulators (Sttover et al 2000). One of these regulatory systems, which is also present in many other bacteria, not only responding to environmental conditions, but mainly response to changes in a bacterial population. This regulatory system detects when it has reached a critical mass of bacteria which has been termed quorum sensing (Fuqua and Greenberg 1998; Fuqua et al 1996, Salmond et al 1995). The mechanism by which gene expression is regulated at high cell densities is based on constant low production per cell of diffusible compounds called autoinducers. When the cell number reaches a certain threshold at which the concentration of autoinducer is sufficiently high, this compound interacts and activates a transcriptional activator which, in turn, changes the pattern of gene expression (Fuqua and Greenberg 1998; Fuqua et al 1996, Salmond et al 1995). In many gram-negative bacteria, the autoinducer are acylated homoserine lactones.

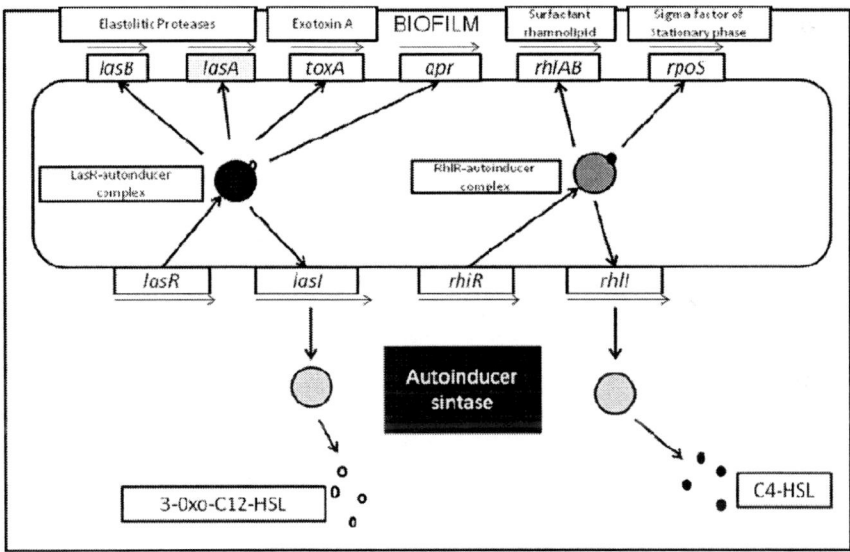

Figure 10. Quorum sensing regulation in *P. aeruginosa*.

P. aeruginosa contains more than one system of "quorum sensing" (Fuqua and Greenberg 1998; Fuqua et al 1996), two of these so-called Las and Rhl systems (Figure 10) have been well characterized at the molecular level. It is known that Las and Rhl systems have a central role in the expression of genes that are related to the virulence of *P. aeruginosa*.

These two regulatory systems interact with each other in a very complex manner to promote the expression of genes involved in the pathogenesis of *P. aeruginosa* (Lam et al 1980, Pearson et al 1997; Pesci and Iglewski 1997, Pesci et al 1997). The system "quorum sensing" based on the LasR transcriptional activator promotes the transcription of several proteases involved in virulence such as the elastases A and B and alkaline protease and exotoxin S.

The autoinducer which interacts with the LasR is N-(3-oxododecanoyl)-homoserine lactone, or PAI1, which is synthesized by the reaction catalyzed by LasI autoinducer synthase enzyme (Figure 10) (Ochsner et al 1994, Ochsner and Koch 1994).

The genes regulates by this quorum sensing system, just depends on cell density and not the composition of the culture medium, so it is assumed to be always expressed that achieves a high density of *P. aeruginosa*, as in the case of any infection.

RhlR protein promotes transcription of genes that encode rhlAB, one of the enzymes that produce the biosurfactant rhamnolipid (Brint et al 1995, Ochsner et al 1994, Ochsner and Koch 1994). It is reported that other metabolites are also involved in the virulence of the bacteria, such as pyocyanin (Britigan et al 1992, Cox 1986), are also under the control of the Rhl system. The autoinducer N-(butanoyl)-homoserine lactone or PAI2 interacts with the RhlR, (Figure 10) produced by the autoinducer synthase RhlI. The transcription of rhlR gene depends on activation of LasR-PAI1, so that the two systems quorum sensing form a regulatory cascade headed by LasR (Lam et al 1980, Pearson et al 1997; Pesci and Iglewski 1997; Pesci et al 1997).

The Rhl system, unlike the Las system, is expressed only in terms of limitation of certain nutrients such as phosphate and nitrogen, so that in a rich medium, even at high cell densities are not expressed factors regulated by RhlR (Pearson et al 1997). Not yet known regulatory systems that determine the differential expression of the Rhl system depending on the availability of nutrients.

Recently it was shown that Las system has a central role in the formation of biofilms, since lasI mutant do not produce these structures, except in the case in which the autoinducer (PAI1) is added (Davies et al 1998). It was also reported by analyzing the genome sequence of PAO1strain a third transcriptional regulator family of quorum sensing activators response called QscR. (Chugani et al 2001). The third controller is called PhzR that is tied to *phz* genes responsible for the synthesis of pyocyanin.

Based on these previous studies we propose that *L. plantarum* be able to make quorum quenching (inhibition of quorum sensing) and thereby inhibit biofilm formation. We set out to examine whether *L. plantarum* does this through enzymes (acylases, lactonases) to destroy autoinducers, or competition through molecules similar to acyl homoserine lactones that can compete for its receptor (LasR or RhlR) in the cytoplasm of *P. aeruginosa*.

IV) *EFFECT OF* HEAT AND PROTEASES ON SLP

To discover whether the inhibitory effect of biofilm was produced by one disrupter enzyme of any component of the matrix (DNAase, Alginases, proteases) or by enzymes destroying autoinducer molecules (acylases, lactonases), we tried SLp with heat (2 h at 100 ° C) and various proteases to see if it retains its inhibitory power. The supernatants treated this way were used for biofilm inhibition assays in plates of 96 wells as discussed above. The results can be seen in figure 11. The heat treated SLp lost virtually all its power inhibitor, while those treated with proteases paradoxically increased inhibition. This would indicate that inhibition is not caused by any type of enzyme.

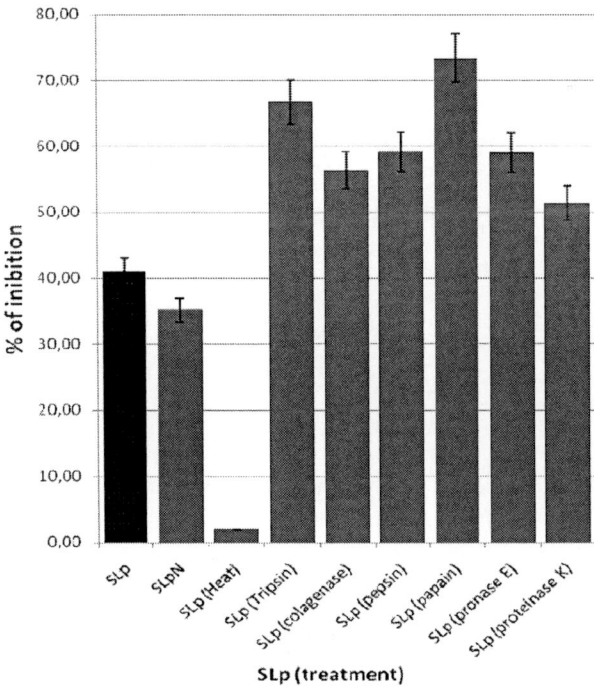

Figure 11. Effect of heat and proteases on the SLp and its impact on power biofilm inhibitor.

To explain why increase the SLp inhibitor power with proteases treatment, we performed the same test with the same protease solutions used before. As shown in figure 12, proteases inhibit biofilm by themselves. This indicates that the increase in inhibitory power in figure 11 is simply the sum of the inhibition produced by the SLp and the respective proteases. For all this we can say then that the inhibition produced by the SLp is not a consequence of the presence of any enzyme.

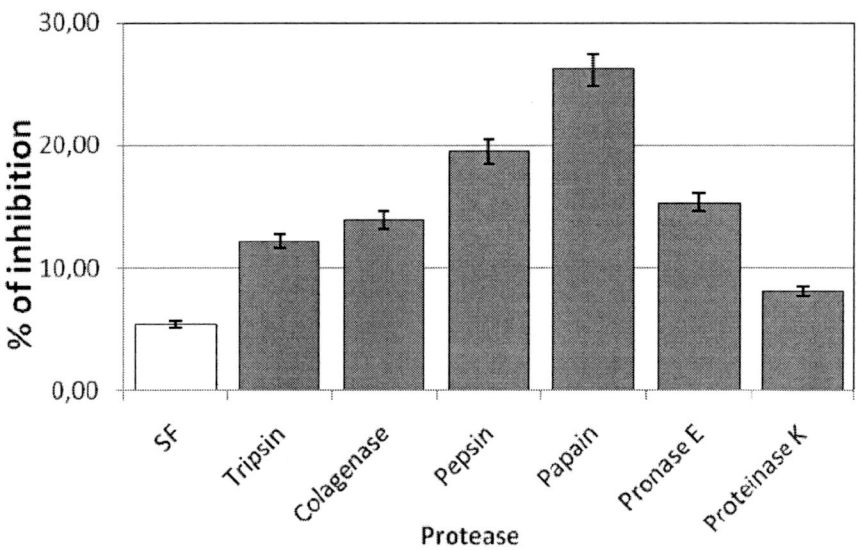

Figure 12. Biofilm inhibition by several proteases.

Effect on Quorum Sensing Molecules

The culture supernatants (Spa and SLp) were extracted two times, with dichloromethane, and each extraction was pooled and taken to complete dryness. In the extract of *P. aeruginosa* supernatant (ESPa) should be present the autoinducers N-butanoyl-homoserine lactones and N-3-oxo-dodecanoyl-homoserine lactones, while in *L. plantarum* (ESLp) as this is gram (+) should be absent. On the other hand, equal volumes of SPa and SLp were incubated enough time (48 h at 37 ° C), to see if SLp had some effect on *P. aeruginosa* signal molecules. After, this mixture was extracted two times, with dichloromethane (EM). Finally the extracts ESPA, ESLp and EM were used to perform the following tests:

Bioassays

Agrobacterium tumefaciens KYC55 (Detect all types of acyl-homoserine lactones). (Zhu et al 2003).

Cromobacterium violaceum 026 (Detect short chain acyl-homoserine lactones C4, C6 and C8) (Llamas et al 2005).

Cromobacterium violaceum vir07 (Detect long chain acyl-homoserine lactones C8, C10, C12 and C14). (Morohoshi et al 2008).

Thin-Layer Chromatography (TLC)

Samples, in volumes of 1–4 ml, were applied to C18 reversed-phase TLC plates (200 mm layer; Baker) and the chromatograms were developed with methanol/water (60:40, vol/vol). After development, the solvent was evaporated, and the dried plates were overlaid with a culture of the indicator bacterium prepared as follows. For 20320 cm plates, a 5-ml overnight culture of *Agrobacterium tumefaciens KYC55* was used to inoculate 50 ml of ABMand the new culture was grown to late exponential phase. The entire 50 ml of culture was added to 100 ml of the same medium containing 1.12 g of melted agar and 60 mg/ml 5-bromo-4-chloro-3-indolyl b-D-galactopyranoside (X-Gal) maintained at 45°C. The culture was mixed thoroughly and immediately spread over the surface of the developed plate held in a Plexiglas jig designed to produce a uniform layer of agar about 3 mm thick. After the agar solidified, the coated plates were incubated at 28°C for 12–18 hr in a closed plastic container.

Gas Chromatography and Mass Spectrometry (CGMS)

To detect chemically the presence or absence of lactones were performed in the extracts, gas chromatography followed by the identification of autoinducer molecules by mass spectrometry. (Hong-Sik Moon 2004).

Fourier Transforms Infrared Spectrometry (FT-IR)

As a chemical complementary method for identifying presence of lactones in the extracts, we used FT-IR through the presence or absence of characteristic peaks corresponding to the lactone carbonyl (Chhabra et al 2003).

Table 1. Identification of lactones in the extracts by different biological and chemical methods

	Extract of *P. aeruginosa* supernatant		Extract of superntants mixture		Extract of *L. plantarum* supernatant	
	C4-HSL	C12-HSL	C4-HSL	C12-HSL	C4-HSL	C12-HSL
Bioassay with A. tumefaciens	+	+	+	+	-	-
Bioassay with C. violaceum 026	+	-	+	-	-	-
Bioassay with C. violaceum vir 07	-	+	-	+	-	-
TLC	+	+	+	+	-	-
GC-MS	+	+	+	+	-	-
FT-IR	+	+	+	+	-	-

For all the above methods we found that homoserine lactones of *P. aeruginosa* are present both in the ESPa and the ME and obviously absent in the ESLp (Table 1). With this compendium of results, we concluded that the supernatant of *L. plantarum* has no direct visible action, detectable by these methods on the lactones of *P. aeruginosa*.

However, with all these methods we could only detect the presence or absence of lactones. Because of this, we do another bioassay with reporter mutant strain *P. aeruginosa* 129b (a generous gift from E. P. Greenberg and K. Lee, University of Iowa, USA). This mutant strain produces neither 3O-C12-HSL nor C4-HSL, but responds to the presence of these compounds by increasing expression of b-galactosidase. It was quantified by Miller reaction (Whiteley et al 1999; Miller JH. 1992). In the direct assay (Figure 13) stimulate the reporter strain with the supernatant of *P. aeruginosa* mixed in equal parts with LB, SLp, SLpN, MRS and a whole-cell culture containing 10^6 *L. plantarum* CFU/mL (T preparation) . In the indirect test (Figure 14), a culture of *P. aeruginosa* was centrifuged, washed with PBS and bacteria were re-suspended in the different samples above. As seen in both trials, β-galactosidase activity is decreased in the presence of SLp, to a lesser extent in the presence of SLpN and can even observe a slight inhibition in the presence of culture medium used to obtain the supernatant (MRS).

These results indicate that the inhibition of lactones activity may not be related with the presence of lactones destroying enzymes.

Because *L. plantarum* has the lux S gene (Sun et al 2004) (Figure 14) essential for the synthesis of type 2 autoinducer (AI-2), and that the phenotypic expression of *P. aeruginosa* can be modified by AI-2 analogues (Ganin et al 2009, Duan et al 2003), we investigated the possibility that AI-2 would be present in SLp, which allow interspecies communication between both bacteria.

Figure 15 show the participation of LuxS enzyme in the synthesis of AI-2 known so far (R-TMF for *Salmonella typhimurium* and *Escherichia coli* and other S-THMF-borate for *Vibrio* spp.)

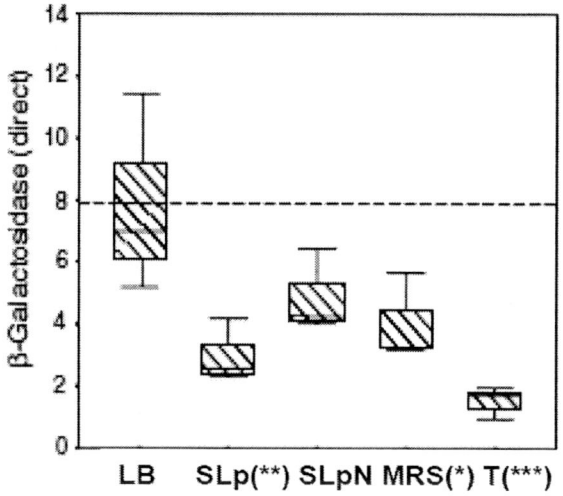

Figure 13. Direct bioassay with the reporter strain *P. aeruginosa* 129b.

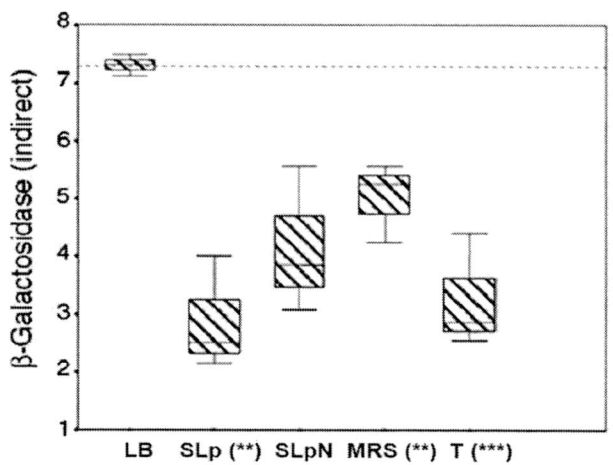

Figure 14. Indirect bioassay with the reporter strain *P. aeruginosa* 129b.

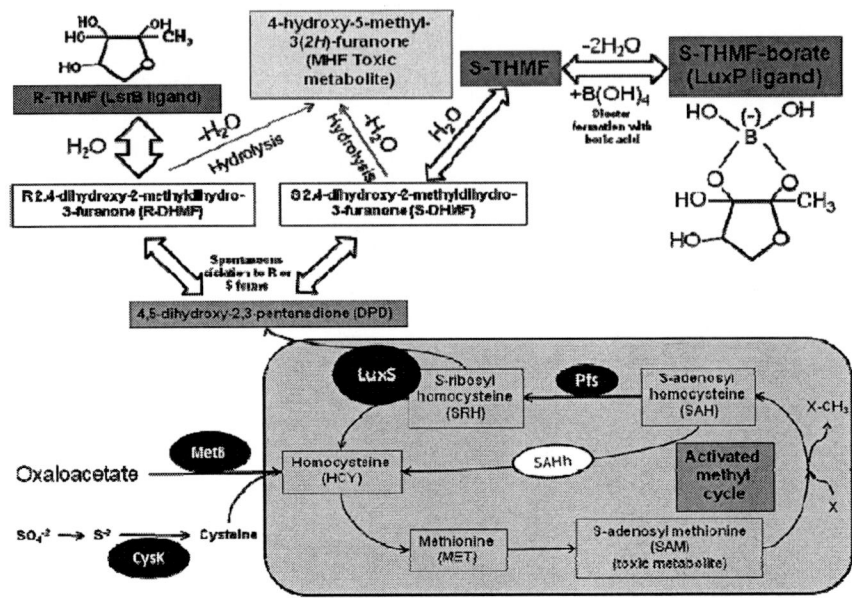

Figure 15. Chemical interconversions of molecules in relation to the activated methyl cycle. The reactions that generate homocysteine (HCY) from (i) sulphate (SO_4^{2-}) through sulphide (S^{2-}), and (ii) oxaloacetate (OA), plus some of the enzymes involved (MetB and CysK) are shown feeding into the boxed reactions that comprise the activated methyl cycle (shaded in light grey). The latter involves the formation of methione (MET) and the subsequent conversion to S-adenosylmethionine (SAM). The activated methyl group (CH_3) of SAM is used for methylation of RNA, DNA, certain metabolites and proteins (X), leading to the formation of the toxic metabolite S-adenosylhomocysteine (SAH). SAH is then removed and the cycle completed by one of two routes, depending on the organism. One route involves the one-step conversion of SAH to HCY by SAH hydrolase (SahH), the other requires the production of S-ribosylhomocysteine (SRH) by Pfs and then the generation of HCY from this by LuxS, which simultaneously generates 4,5-dihydroxy-2,3-pentanedione (DPD). There is spontaneous cyclization of DPD to either the *R* or *S* form of 2,4-dihydroxy-2-methyldihydro-3-furanone (DHMF). *R*-DHMF can then undergo hydration to form the molecule which was co-crystalized with LsrB, (2*R*,4*S*)-2-methyl-2,3,3,4-tetrahydroxytetrahydrofuran (*R*-THMF, box). Alternatively, hydrolysis can occur to create 4-hydroxy-5-methyl-3(2*H*)-furanone (MHF, box). MHF can also be formed through hydrolysis of *S*-DHMF, which can additionally undergo hydration to form *S*-THMF and subsequently form a diester with boric acid to generate the ligand found in complex with LuxP (dark grey box). Where known, the reversible nature of interconversions are indicated by the double arrows. Based on data from Winzer et al 2003; Miller et al 1968 and Nedvidek et al 1992.

To check the presence of AI-2 in the SLp, we performed a semi-quantitative bioassay with *V. harveyi* BB170. This reporter strain responds to the presence of AI-2 by producing bioluminescence (Sigrid et al 2003). The test was based on stimulation of the reporter strain with supernatants of *V. harveyi* BB120 (positive control for AI-2) which considered 100% stimulation. Glucose causes metabolic inhibition to *V. harveyi* BB170 (Turovskiy et al 2006). For this we grow *L. plantarum* in modified MRS (glucose was replaced by galactose). The AI-2, which is chemically furanones, can be borates or not by what is known so far (Figure 15). Because of this we grow *L plantarum* in another modified MRS (MRS with the addition of boric acid). The results obtained (Figure 16) shows that the acidity of the supernatants inhibit the growth of reporter strain therefore does not stimulate bioluminescence. In contrast, the supernatants neutralized exhibit more than 10% of stimulation, which according to Bassler (Bassler et al 1997) is sufficient to say that there are AI-2 activity in the sample. Because the induction was higher with supernatant contained boric acid, we can also say that the type of AI-2 present in the sample would be a borate-furanone-type like that produced by *Vibrio* spp (Figure 15), but still remain check this for chemical methods.

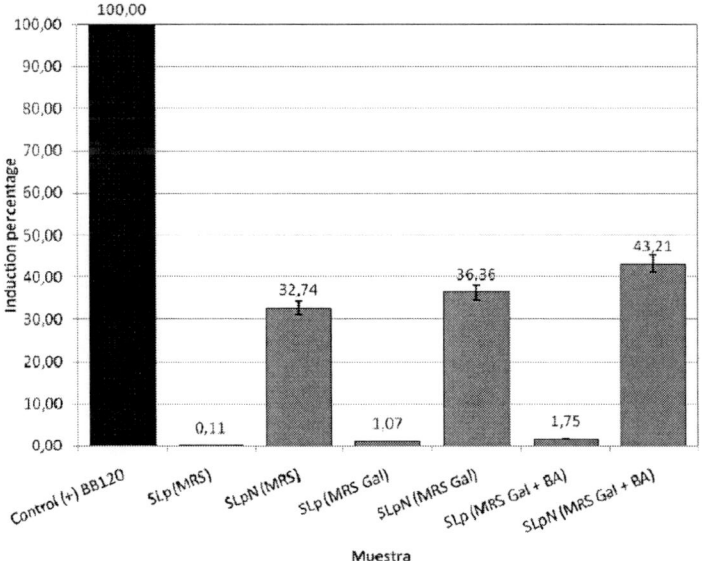

Figure 16. *Vibrio harveyi* BB170 reporter strain stimulation with different *L. plantarum* supernatants.

However, there not have been described in *P. aeruginosa* receptors for type 2 autoinducer known so far (Figure 17), so it might seem that the AI-2 molecule found in the SLp would have a competitive activity for HSL of *P. aeruginosa* as with the halogenated *Delisea pulcra* Furanones (Hentze et al 2003; Manefield et al 2002)

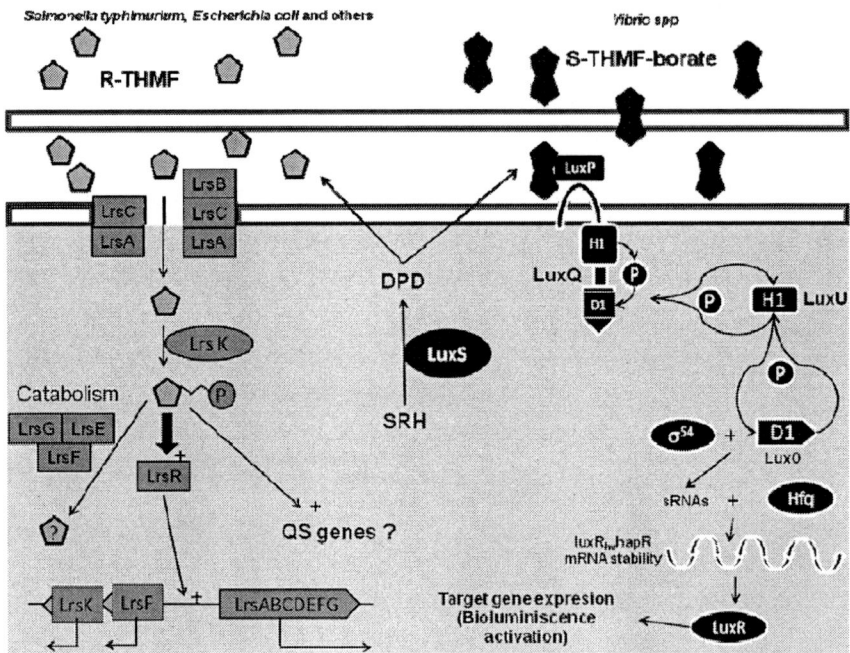

Figure 17. LuxS (depicted as a black oval) converts *S*-ribosylhomocysteine (SRH) to the precursor of autoinducer 2 (AI-2), 4,5,-dihydroxy-2,3-pentanedione (DPD), in the cytoplasm. DPD then undergoes spontaneous cyclization and export to the culture supernatant. Depending on the bacteria, response to AI-2 can follow one of the two currently identified routes. In one group of bacteria exemplified by *Salmonella* (shown on the left), the cyclic derivative of DPD, (2R,4S)-2-methyl-2,3,3,4-tetrahydroxytetrahydrofuran (*R*-THMF, grey pentagons) is found bound to a homologue of the periplasmic binding protein LsrB. LsrB is part of an ABC transporter (encoded by the *lsrACDBFGE* operon). The putative ATPase of the ABC transporter (LsrA), sugar binding protein (LsrB), heterodimeric membrane channel (LsrC and LsrD), LsrF and LsrG show similarity to proteins encoded by the *b1513* operon in *Escherichia coli* 45, and similar operons exist elsewhere. This operon is regulated by the repressor LsrR, the gene of which is located upstream. Immediately downstream of *lsrR* is a gene (*lsrK*) with similarity to the xylulokinase gene (*xylB*). LsrK phosphorylates AI-2, probably following import to

sequester it in the cytoplasm. Phosphorylated AI-2 then causes LsrR to relieve its repression of the *lsr* operon, allowing further AI-2 import. LsrF and LsrG are necessary for further processing of phospho-AI-2 and form a product(s) no longer capable of preventing the repression by LsrR. It is not clear whether LsrF (which resembles an aldolase), or LsrG (unknown function) act on different substrates (for example, phosphates at different positions of AI-2, or chemically distinct forms of AI-2), or whether they catalyse reversible reactions that can occur spontaneously at a low level. It is however clear that both are required for the further processing of phospho-AI-2. The product of the gene with similarity to ribulose phosphate epimerase (LsrE) is thought to contribute to this conversion, but its function is unclear. The alternative pathway of AI-2 response described for *Vibrio* spp. involves a phosphorelay signalling transduction (shown on the right). The form of AI-2 that is active in this system is the furanosyl-borate-diester (*S*-THMF-borate: black double pentagons) in complex with the periplasmic binding protein, LuxP. The current model suggests that LuxP then interacts with the inner membrane protein, LuxQ, inducing a conformational change that confers a phosphatase activity in LuxQ. This process extracts a phosphate from a two component phosphorelay protein (LuxU) that in turn dephosphorylates the response regulator, LuxO, leading to activation of bioluminescence. At low cell densities (in the absence of AI-2), the proteins involved are converted into kinases, which reverses the flow of phosphates to create phosphorylated LuxO. Phosphorylated LuxO acts alongside σ54 to activate the production of multiple, redundant small regulatory RNAs (sRNAs) which interact with the mRNA of the transcriptional activator LuxRVh (HapR in *Vibrio cholerae*), causing Hfq-dependent destabilization. As LuxRVh activates transcription of the *luxCDABEGH* operon (which encodes the enzymes required for light generation), the reduced levels of LuxRVh protein in the cells leads to a reduction in bioluminescence. As the dephosphorylated LuxO generated by the presence of AI-2 (at high cell densities) cannot activate the production of sRNAs, the *luxRVh* mRNA has an increased stability, and the resultant LuxRVh induces bioluminescence by activating the *luxCDABEGH* operon. The phosphorelay is depicted by the arrowed circles containing 'P', and the direction of transfer in the presence of AI-2 is indicated by the solid black arrows. Figure based on data from **Lenz et al 2004 and Taga et al 2003**.

IV) INHIBITION OF VIRULENCE FACTORS (ELASTASE)

Elastase Assay

The elastolytic activity of the different samples was investigated after 1 h and 18 h, using elastin Congo red (Sigma, St Louis, MO, USA) as a substrate (Gambello et al 1991). Insoluble elastin Congo red was removed by centrifugation, and the absorbance of the supernatants was measured at 495 nm. For the 1-h assay, *P. aeruginosa* at OD600 0.7 was washed and resuspended in fresh LB medium, and then diluted 1:7 in T, SLp, SLpN, MRS broth and a whole-cell culture containing 10^6 *L. plantarum* CFU/mL (T preparation). After incubation for 1 h, the elastolytic activity of the samples was determined. For the 18-h assay, a 16-h *P. aeruginosa* culture was mixed with the *L. plantarum* samples and MRS broth, as described above, and elastolytic activity was determined after incubation for 18 h.

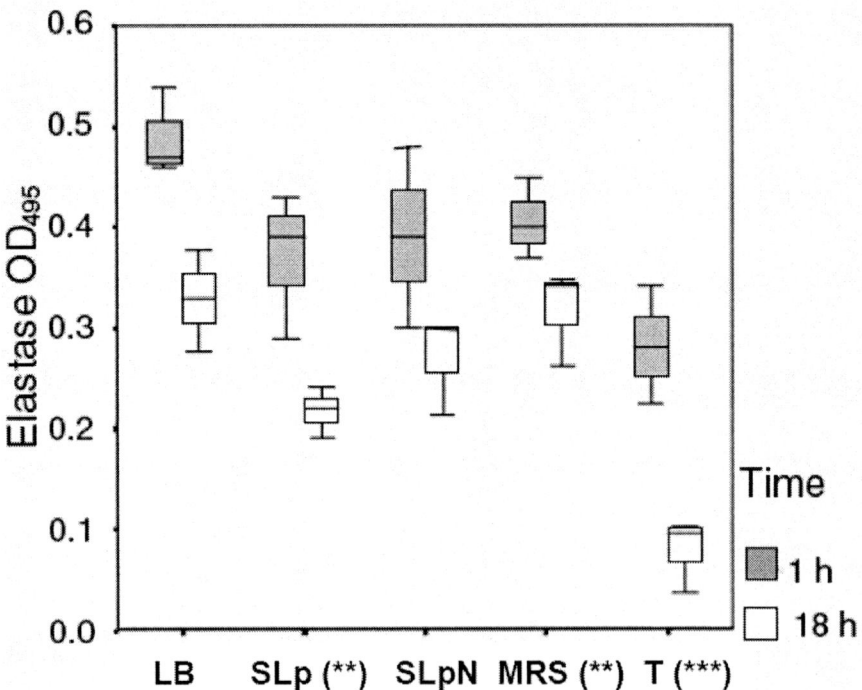

Figure 18. Inhibitory effect of different preparations of *Lactobacillus plantarum* on elastase produced by Pseudomonas aeruginosa. The solid line inside each box is a mean of triplicate samples. LB, *P. aeruginosa* in Luria–Bertani medium; SLp, *P. aeruginosa* in the presence of *L. plantarum* acid filtrate; SLpN, *P. aeruginosa* in the presence of *L. plantarum* neutralized filtrate; MRS, *P. aeruginosa* in MRS broth; T, *P. aeruginosa* in the presence of 10^6 CFU L. plantarum / mL. *p < 0.05; **p < 0.01; ***p < 0.001.

Figure 18 shows that elastolytic activity was inhibited significantly by the T (p<0.001), SLp (p<0.01) and SLpN (p<0.05) preparations compared with *P. aeruginosa* cultured in LB medium.

The *P. aeruginosa* quorum-sensing signals are important for the control of elastase, rhamnolipid and the formation of differentiated biofilm (Smith et al 2002; Sandoz et al 2007). The inhibitory effect shown in the present study may be associated with the inhibition of quorum-sensing molecules, but a direct inhibitory action of *L. plantarum* secondary metabolites on elastase and biofilm formation cannot be ruled out. The acid *L. plantarum* growth medium itself had some inhibitory activity, but the greatest inhibitory effect was observed with the T and SLp preparations.

All this leads us to believe that the mechanism of inhibition of biofilm of *P. aeruginosa* with SLp is a sum of several factors:

Lactic Acid

Lactic acid can be considered to be a key antimicrobial compound produced by lactobacilli (Servin et al 2004). For example, the strong antimicrobial activity of *L. rhamnosus* GG against *S. enterica* serovar *Typhimurium* was shown to be due to the accumulation of lactic acid (De Keersmaecker et al 2006). The exact mode of action underlying this observed antimicrobial effect of lactic acid has not yet been completely clarified, although it is clear that both *Salmonella* growth and the expression of virulence factors are affected by lactic acid (Durant et al 2000). Besides exerting its activity through lowering the pH and through its undissociated form, lactic acid is also known to function as a permeabilizer of the gram-negative bacterial outer membrane (Alakomi et al 2000), allowing other compounds to act synergistically with lactic acid. In addition, organic acids such as lactic acid can capture elements essential for growth, such as iron, by their chelating properties (Presser et al 1997). Furthermore, we have also found that lactic acid has a direct effect on the viability of *P. aeruginosa* in both planktonic and biofilm phase (Figures 8 and 9).

Bacteriocins

L. plantarum can produce bacteriocins called plantaricin. These molecules are usually active against closely related bacteria that are likely to reside in the same ecological niche. These bacteriocins (class II bacteriocins) act generally by inducing membrane permeabilization and the subsequent leakage of molecules from target bacteria (Eijsink et al 2002). In *L. plantarum*, bacteriocin production is controlled in a population density-dependent manner using a secreted peptide pheromone for quorum sensing (QS). The R-IVET study of *L. plantarum* WCFS1 already gave some indications, as a *plnI* gene, encoding the plantaricin immunity protein, was found to be induced in the murine GIT (Bron et al 2004). This gene belongs to a bacteriocin locus in *L. plantarum* WCFS1 that includes *plnABCD*, encoding a plantaricin 2CRS and an autoinducing peptide, PlnA, and several genes encoding class II bacteriocins (*plnE-plnF*, *plnJ-plnK*, and *plnN*) (Kleerebezem et al 2003; Sturme et al 2007). Marco further confirmed that *plnI* is induced in the small intestine, cecum, and colon of mice (Marco et al 2007), suggesting that the production of this bacteriocin is important for *L. plantarum* in the highly competitive environment of the gastrointestinal tract (GIT). However, no mutant analysis has yet been performed to confirm this hypothesis.

Lactobacillus plantarum LP 31 strain produces an antimicrobial compound that inhibits the growth of food-borne pathogenic bacteria. It was inactivated by proteolytic enzymes, was stable to heat and catalase and exhibited maximum activity in the pH range from 5.0 to 6.0. Consequently, it was characterized as a bacteriocin. It was purified by RP (reverse-phase) solid-phase extraction, gel filtration chromatography and RP-HPLC. Plantaricin produced by *L. plantarum* LP 31 is a peptide with a molecular weight of 1558.85 Da as determined by MALDI-TOF mass spectrometry and contains 14 amino acid residues. It was shown to have a bactericidal effect against *Pseudomonas* sp, *Staphylococcus aureus*, *Bacillus cereus* and *Listeria monocytogenes* (Muller et al 2009).

H_2O_2: H_2O_2 production by lactobacilli has also been suggested to be an important antimicrobial mechanism, especially in the vagina of healthy women (Servin et al 2004).

Competitive Exclusion

Intestinal pathogens such as type 1-fimbriated *E. coli* utilize oligosaccharide receptor sites in the gut (Le Bouguenec 2005). There is some evidence that probiotics could use the same attachment site so that the pathogen is in competition for binding to the host mucosal interface and thereby could be inhibited from invading the mucosal layer. This antipathogenic mechanism is known as competitive exclusion and generally requires that the probiotic lactobacilli are administered in a preventive setup, as the displacement of a pathogen by a *Lactobacillus* strain is usually not observed. The adhesins described above probably contribute to this mechanism of probiotic action, although a specific mechanisms based on steric hindrance are also possible. One putative competitive exclusion factor is the mannosespecific adhesin Msa of *L. plantarum* (Pretzer et al 2005). Its identification allows detailed studies with deletion and overproduction strains to identify strains that effectively exclude pathogens containing type 1 fimbriae. Mack et al demonstrated that a spontaneous and non-characterized *adh* mutant of the probiotic strain *L. plantarum* 299v, but which is suggested to be affected in the *msa* gene, showed 10-times-less adherence to HT-29 epithelial cells and was not able to inhibit the adherence of enteropathogenic *E. coli*, in contrast to the wild type (Mack et al 2003). In our specific case we have ascertained that the SLp is able to decrease the adhesion of *P. aeruginosa* perhaps by a mechanism related to acidity (see Figure 5).

Bacterial Cell-Cell Communication

Given the high number and level of diversity of bacteria that comprise the GIT, it was postulated that the members of this community somehow communicate to coordinate various adaptive processes that include competition and cooperation for nutrients and adhesion sites (Kaper et al 2005). QS is a cell-to-cell signaling mechanism through which bacteria produce and/or respond to chemical signals called autoinducers.

Interspecies Communication

In the late 1990s, a new family of signal molecules, autoinducer-2 (AI-2), and its cognate synthase LuxS, which are present in both gram-negative and gram-positive bacteria, were described (Surette et al 1999). In *Vibrio harveyi*, AI-2, a furanosyl borate diester, is one of the signals that regulate bioluminescence through a complex phosphorelay system (Henke et al 2004). The binding of AI-2 to the periplasmic receptor LuxP modulates the activity of the inner membrane sensor kinase LuxQ, transducing AI-2 information into the cytoplasm. *V. harveyi* can produce light in response to the AI-2 produced by many other bacterial species. These observations of *V. harveyi* resulted in the development of a bioassay to detect AI-2 production and led to the suggestion that AI-2 acts as a universal signal molecule that fosters interspecies cell-cell communication (Bassler et al 1997). Using the *Vibrio* bioluminescence assay, De Keersmaecker and Vanderleyden tested the spent culture supernatant of various lactobacilli for the presence of AI-2- type molecules. Strains such as *L. rhamnosus* GG, *L. plantarum* NCIMB8826, *L. johnsonii* VPI1088, and *L. casei* ATCC 393 were shown to produce AI-2-like molecules (De Keersmaecker et al 2003). In the gastrointestinal pathogen *Vibrio cholerae*, the accumulation of two QS signals, cholerae AI-1 (CAI-1) and AI-2, represses the expression of virulence factors at high population density. AI-2 functions synergistically with CAI-1 to control virulence gene expression, although CAI-1 is the stronger of the two signals (Higgins et al 2007). However, apart from bioluminescence in *V. harveyi* and a role in virulence of *V. cholerae*, no obvious phenotype has been associated with the extracellular accumulation of this molecule in most bacteria (Vendeville et al 2005). The functional studies of AI-2-mediated QS are complicated by the fact that the AI-2 synthase LuxS forms an integral part of the activated methyl cycle, as mentioned above in relation to stress resistance. In this cycle, LuxS catalyzes the conversion of *S*-ribosylhomocysteine, yielding (*S*)-4,5-dihydroxy-2,3-pentanedione and homocysteine (Vendeville et al 2005). (*S*)-4,5-Dihydroxy-2,3-pentanedione undergoes spontaneous rearrangements to form AI-2, while homocysteine is recycled to methionine and SAM. The disruption of *luxS* also disrupts the activated methyl cycle, resulting in many metabolic defects that cannot be easily discriminated from QS defects. However, since only the LuxS AI-2 synthase is widespread in bacteria, as opposed to the LuxPQ AI-2 receptor, most functional studies have been performed with *luxS* knockout mutants. This is also the case for lactobacilli, and only two functional analyses of *luxS* knockout mutants have been described in detail. In *L. reuteri* 100-23, the disruption of *luxS* resulted in increased biofilm formation in vitro in a bioreactor and on the epithelial surface of the murine forestomach (Tannock et al 2005). Whether this was due to disrupted QS control of the biofilm thickness or disturbed metabolism is difficult to discriminate, as the addition of purified AI-2 to the biofilm culture medium could not restore the phenotype to the wild-type level, and the *luxS* mutant showed a reduced ATP level in exponentially growing cells. Moreover, the ecological performance of the *luxS* mutant, when in competition with *L. reuteri* strain 100-93, was significantly reduced in the highly competitive conditions of the murine cecum but not in the stomach or jejunum (Tannock et al 2005). In *L. rhamnosus* GG, the disruption of the *luxS* gene resulted in pleiotropic effects on in vitro growth, biofilm formation, and in vivo persistence in the murine GIT, and these effects were shown to be caused merely by metabolic defects (Lebeer et al 2008; Lebeer et al 2007). Therefore, although AI-2 is an attractive candidate for multispecies communication in the GIT, it is very difficult to verify this hypothesis in vivo. Clearly, the strategy to investigate this hypothesis has to include, in addition to the characterization of *luxS* mutants, the identification of putative AI-2 receptors and signaling pathways between different species, both pathogenic and probiotic bacteria.

Chapter 4

EX VIVO ASSAYS

Multiple factors such as infection, ischemia, advanced age, malnutrition and diabetes have been identified as contributors to impaired wound healing (Fonder et al 2008). With regard to infection, pathogenic bacteria delay wound healing through several different mechanisms such as persistent production of inflammatory mediators, metabolic wastes and toxins, or maintenance of necrotic neutrophils, which release cytolytic enzymes and free oxygen radicals. This prolonged inflammatory response contributes to host injury and delays healing. The need for novel antimicrobial agents to combat evolving antibiotic resistance among human bacterial pathogens is clear and imminent (Bjarnsholt et al 2008). Unlike antibiotics that act selectively on a specific target, antiseptics have multiple targets and a broader activity spectrum. The strongest argument against the application of antiseptics to wounds is that they are cytotoxic to cells such as fibroblasts, keratinocytes and leukocytes that are essential for the wound healing process (Cooper et al 1991). One of most frequent pathogens isolated from chronic infections is *P. aeruginosa*, a Gram-negative opportunist (Kievit et al 2000). Its resistance to antimicrobial agents and numerous virulence factors means that *P. aeruginosa* causes many recalcitrant infections (Lewis 2001).

Previously in this chapter we show that *L. plantarum* interferes with the pathogenic capacity of *P. aeruginosa* (quorum sensing, biofilm, virulence factors and growth (Valdez et al 2005; Ramos et al 2008). Also, we explained that *L. plantaum* have antagonic effect than pathogens on the immune cells. Lactobacilli induce a cytokine pattern in peripheral blood cells which is completely different from *P. aeruginosa* (Hessle et al 2000). Because a topical treatment with *L. plantarum* cultures is currently being carried out by our medical team with infected burns and chronic venous ulcers in humans (Peral et al 2009, Peral et al 2009b) with encouraging results, we investigate the effect of *L. plantarum* supernatant on PMN, the more constant cell involved in the infected chronic wound. The aim of our work was to compare in vitro polymorphonuclear (PMN) damage caused by *L. plantarum* and *P. aeruginosa* supernatants and mixtures of both as well as acetic acid. Acetic acid is used as an antiseptic agent against *P. aeruginosa* (Sloss et al 1993) and is frequently used in our hospitals.

We determine the intracelular pH, apoptosis necrosis and viability of PMN incubates with different supernatants (Ramos et al 2010).

BACTERIAL STRAINS

Two *P. aeruginosa* strains were used in this study: a standard clinical isolate (PaC) and a qsc mutant (Pa129b). This mutant strain is a non producer of AHL. The *L. plantarum* strain used was ATCC 10241 (Lp).

Supernatants

PaC was grown for 12 h at 37 °C in Luria–Bertani (LB) medium (Gibco, Rockville, MD, USA). Pa129b was grown in LB medium containing 100 µg/ml gentamicin and then for 12 h in LB at 37 °C. *L. plantarum* ATCC 10241 was grown for 12 h in MRS broth (Britania, Buenos Aires, Argentina) at 37 °C. Supernatants of PaC, Pa129b and Lp were recovered after centrifugation and subsequent filtration through 0.22 µm filters and the pH was determined: SPaC (pH 6.33±0.27), S129b (pH 6.99±0.32) and SLp (acid filtrate, pH 5.22±0.43). Aliquots of SLp were neutralized with 8 M NaOH to pH 7 (SLp neutralized: SLpN).

Supernatant Mixtures

The following supernatant mixtures (vol:vol) were prepared with corresponding pH: SPaC-SLp (pH 5.54±0.35), SPaC-SLpN (pH 6.71±0.25), S129b-SLp (pH 5.48±0.37) and S129b-SLpN (pH 7.00±0.30).

IN VITRO ASSAYS TO DETECT ENDOTOXINS (LPS) IN SLpN AND MRS MEDIUM

LPS is the major component of the outer membrane of Gram-negative bacteria. LPS is an endotoxin, and induces a strong response from immune systems. LPS acts as the prototypical endotoxin because it binds the CD14/TLR4/MD2 receptor complex, which promotes the secretion of pro-inflammatory cytokines in many cell types, but especially in macrophages. LPS is an exogenous pyrogen (external fever-inducing substance) and must be tested in all preparations for any human treatment.

The ELISA was performed following standard assay development procedures described elsewhere (Crowther 2001). Brefly, LPS from Escherichia coli O56:B6 (Sigma, St. Louis Mo, U.S.A.) diluted in PBS pH 7.4 to give a range of concentrations from 1 to 1000 ng ml−1, SLpN and MRS medium undiluted and diluted 1:10 and 1:100 in PBS pH 7.4 were prepared and 100 µl of each of these samples were used to coat 96-well ELISA plates (NUNC Maxisorp, Denmark). After blocking with BSA, a dilution of rabbit Polyvalent O antisera was added. These antibodies were detcted with goat anti-rabbit immunoglobulin G specific HRP conjugate (Sigma St. Louis Mo, U.S.A) and peroxidase color substrate (OPDA+H2O2). Absorbance values at a wavelength of 490 nm (A490) were obtained for each sample well.

LPS in SLpN and MRS medium: The Polyvalent O antisera reacted strongly with all dilutions of LPS and did not react with SLpN or MRS medium. The mean A490 values for LPS ranged from 0.230 ± 0.030 (1 ng) to 0.875 ± 0.042 (100 ng) and for SLpN 0.009 ± 0.003 and MRS medium 0.010 ± 0.004, indicating that SLpN and MRS medium were LPS free. The only source of LPS in SLp and MRS medium might come from the yeast extract, which is produced from baker's yeast by autolysis at 50–55 °C or by plasmolysis in the presence of high concentrations of NaCl. Yeast extract contains amino acids, peptides, water soluble vitamins and carbohydrates without components of cell wall. The LPS is a wall component; hence, MRS and SLp should not have LPS as our results indicate.

Isolation of PMNs

Heparinized blood samples were collected by venipuncture in healthy individuals of both sexes and 25 to 35 years of age (n=10). Neutrophils were isolated by dextran T-500 (Sigma, St. Louis MO, USA) sedimentation and Ficoll-Hypaque (Sigma, St. Louis MO, USA) gradient centrifugation. Viability of the PMN population measured with Trypan Blue was >96%. Finally, cells were suspended at a concentration of 1×10^6 PMNs/ml in PBS, pH 7.4.

PMN Suspensions

Neutrophils (1×10^6 cells) were suspended in each supernatant and mixture. For comparison between acetic acid (AA) and SLp effects, PMNs were suspended in both AA and SLp solutions diluted with PBS with the following pH: 5.22, 6.08 and 6.75. After incubation at 37 °C for 30 min (unless indicated otherwise) viability and apoptosis/necrosis were evaluated.

PMN Viability

PMN suspensions were stained with Trypan Blue after incubation.

Intracellular pH of PMNs (pHi)

Determination of pHi was performed using carboxy-SNARF-1-AM as previously described (Bassnett et al 1990). PMN suspensions were washed twice and suspended in 1 ml of PBS, pH 7.4, loaded with 10 µM carboxy-SNARF-1-AM (Molecular Probes). After 15 min of incubation, cells were washed, re-suspended in PBS and analyzed in a Partec Pas II flow cytometer (Partec, GmbH, Munster, Germany) with excitation at 488 nm and emission at 570 nm (FL2) and 620 nm (FL3). Intracellular pH was estimated from the emission intensity ratio at the two wavelengths and standardized by comparison with the fluorescence intensity ratios of cells whose pHi values were fixed by incubation with 10 µM nigericin in high-potassium buffers. Ten thousand events were collected and data were plotted as forward scatter vs fluorescence ratio (size vs. pHi).

Effect of extracellular pH (pHo) on pHi and viability of PMN: When PMNs were incubated with SLp and SLpN, their pHi was correlated with the pHo; SLp (acidic) produced a decrease in cell viability when compared to NSLp (neutral) ($p<0.05$). In contrast, incubation of PMNs with *P. aeruginosa* supernatants, SPaC (pH=6.33±0.27) and S129b (pH=6.99±0.32), induced a remarkable cytoplasmic acidosis showing a lack of correlation between pHo and pHi with a remarkable decrease in viability (p<0.001). The pHi and viability of PMNs incubated with mixture of supernatants (SPaC-SLp, SPaC-SLpN, S129b-SLp and S129b-SLpN) were higher than pHi and viability of PMN incubated with SPaC and SP129b alone (p<0.01) (Figure 19). When comparing the effect produced by SLp and acetic acid on PMN pHi, a correlation between pHi and pHo could be observed. Viability with SLp and AA diminished with decreasing pHo. Incubation with acetic acid produced a higher intracellular acidosis and lesser viability (p<0.05) (Figure 20).

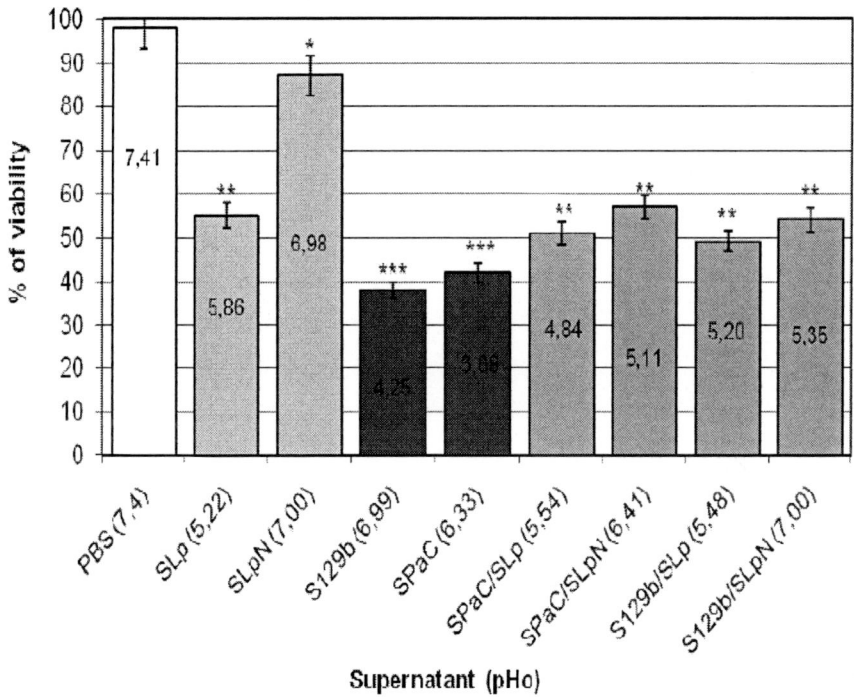

Figure 19. Effect of supernatants and mixtures of supertnatants on viability and intracellular pH (pHi) of neutrophils. The supernatant pH (outer) is indicated as pHo and the intracellular pH (pHi) is indicated inside each bar. The viability was measured by Trypan Blue and pHi by SNARF-1 stains. Significant difference compared to PBS: *$p<0.05$, **$p<0.01$, ***$p<0.001$.

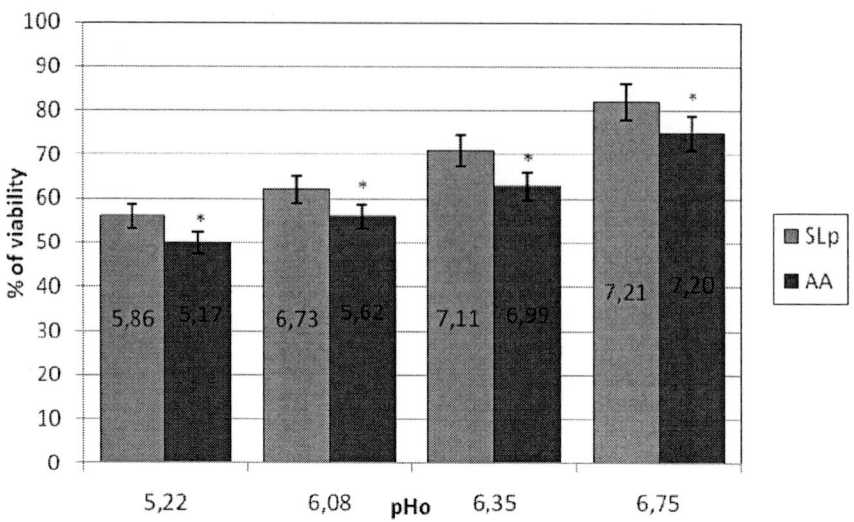

Figure 20. Effect of SLp and AA solutions with different pH (pHo) on viability (Trypan Blue) and intracellular pH (pHi) of neutrophils. The pHi is indicated inside each bar. Significant difference comparing SLp vs AA: * $p < 0.05$.

DETERMINATION OF PMN APOPTOSIS/NECROSIS BY ANNEXIN V AND PROPIDIUM IODIDE (PI)

PMNs (1×10^6 cells) were suspended in 1 ml of different pH solutions of SLp and acid acetic and then incubated at 37 °C for 15, 30 and 45 min. Doxorrubicin (2 µg/ml) was used as apoptosis control. Cells were washed twice with cold phosphatebuffered saline (PBS) and then with 1X binding buffer (10X binding buffer: 0.1 M HEPES in NaOH, pH 7.4; 1.4 MNaCl; 25 mMCaCl2). Next, PMNs were suspended in 100 µL of 1X binding buffer supplemented with 5 µL of Annexin V-FITC solution (Pharmingen, San Diego, CA) and 5 µL of PI solution (50 µg PI/ml PBS, pH 7.4) and suspensions were incubated in the dark at room temperature for 15 min. Within 1 h after addition of 400 µL of binding buffer, samples were analyzed in a Partec Pas II flow cytometer with excitation at 488 nm and emission at 540 nm (FL1) for Annexin V-FITC and 570 nm (FL2) for PI. Ten thousand events were collected and data were classified as follows: Annexin V(+)/PI(+): dead cells; AnnexinV (−)/PI(−): viable cells and Annexin V(+)/PI(−): early apoptotic cells. Data of each population are given as percentage (Vermes et al 1995).

As shown in figure 21, an increase in cell apoptosis and necrosis was observed, dependent on the incubation time and acidity of both SLp and AA. SLp apoptosis/necrosis values were lower than AA after every incubation time and at each pHo.

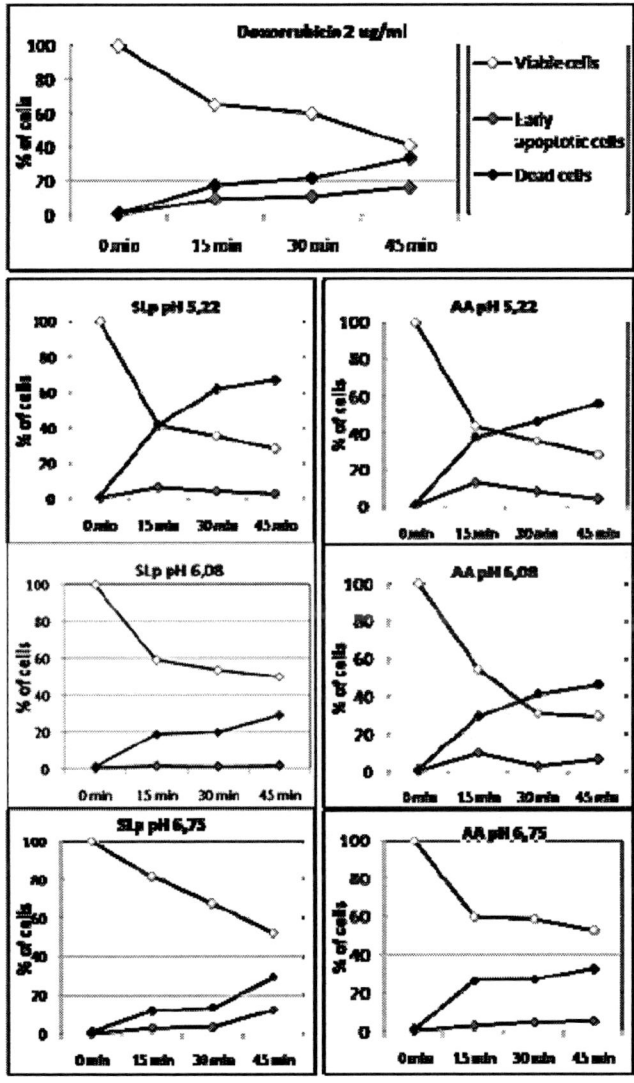

Figure 21. Comparison of the effect of SLp and AA with different pH (pHo) on apoptosis and necrosis of neutrophils after different incubation times.

IN VIVO ASSAYS

Murine Inflammatory Response to Supernatants

To investigate the in vivo inflammatory response induced by the supernatants and supernatant mixtures mentioned above, adult Balb/c mice were inoculated subcutaneously with 0.1 ml of each specimens; a control group was inoculated with a solution of AA (pH 5.22; the same pH as SLp) and PBS. Mice were sacrificed 2 or 24 h after inoculation and blood and skin samples were taken. The skin and connective tissue around the inoculation site were excised, fixed in 10% formalin and embedded in paraffin. Uniformly thin sections (5 µm) were cut and stained with hematoxylin/eosin for light microscopy examination. Inflammation parameters such as vascular congestion, edema, cellular infiltration, necrosis and apoptosis were assessed. Cell infiltration was quantified and classified through microscopy (40X) as follows: Highly Abundant (HA: >100 cells/field); Abundant (A: 50–100 cells/field); Regular (R: 10–50 cells/field) and Weak (W: 1–10 cells/field). Blood collected was used for white blood cell counts using a Neubauer's chamber and Giemsa staining. The protocol used was similar to the one approved by The Animal Care and Use Committee at Texas Tech University Health Sciences Centre.

Inflammatory response in Balb/c mice (Table 2): Perivascular and interstitial inflammatory infiltrates as well as vasodilatation and edema in epidermis and dermis were observed in all skin samples except for controls. Infiltrated cells were mainly PMNs. In all cases necrosis tissue was evidential. The most pronounced PMN infiltration, edema and vasodilatation were observed 2 h after inoculation with SpaC, S129b and AA, when compared to SLp or SLpN. A lower inflammatory response was observed with a mixture of supernatants (SPaC-SLp, SPaC-SLpN, S129b-SLp and S129b-SLpN) compared to inoculation of pure *P. aeruginosa* supernatants (SPaC and S129). After 24 h of inoculation a decrease in cell infiltration was observed compared to 2 h of inoculation.

Table 2. Blood cell (leukocytes) counts and cellular skin infiltration in mice after 2 h of inoculations of PBS, MRS medium, LB medium, *L. plantarum* supernatants, *P. aeruginosa* supernatants, mixtures of supernatants, and acetic acid (see text for details)

	Leucocytes/µl	%Neutrophils	Infiltration
PBS	6570 - 7730	20.6 – 31.2	W
MRS	6200	35	W
LB	6800	39	R/W
SPaC	11158	57	HA
SPa129b	7500	48	A
SLp	6516	39.5	R
NSLp	6950	38	R/W
SPaC+SLp	7275	25.5	A
SPa129b+SLp	5125	52.5	R
SPaC+NSLp	5900	44	A/R
S129b+NSLp	6950	41.5	R
AA	7900	39	A

We showed that *L. plantarum* culture supernatant in vitro inhibited quorum-sensing molecules, biofilm, elastase and growth of *P. aeruginosa*. Acidity was found to be fundamental in this mechanism (Valdez et al 2005; Ramos et al 2008). Acid antiseptics are frequently used in hospitals for the treatment of chronic and infected wounds because they reduce the bacterial load (Sloss et al 1993). However, certain authors disapprove the use of antiseptics in open wounds due to their cytotoxic effect on essential cells in the wound healing process, such as fibroblasts, keratinocytes and leukocytes (Cooper et al 1991). This cytotoxicity seems to be concentration-dependent. Because of this, several antiseptics are used at low concentrations because they are not cytotoxic and retain their antibacterial activity in vitro (Sloss et al 1993). Acetic acid, an antiseptic frequently used in our hospital, seems to have this characteristic, but reports about its cytotoxicity are contradictory (Cooper et al 1991; Sloss et al 1993). Ours results showed that cytotoxicity to PMNs caused by SLp and AA depended of the incubation time and acidity and that, in all cases, the cytotoxicity of SLp was lower than that of AA. *L. plantarum* can produce a mixture of formic, acetic, propionic, butyric and lactic acid. Topical applications to infected wounds of mice and humans lessened the bacterial load and improved the activity of PMNs obtained from lesion tissue by promoting bacterial phagocytosis, modifying production of IL-8 and diminishing their apoptosis/necrosis. Low oxygen tension (Detmar et al 1997) and elevated lactic acid may also stimulate angiogenesis by stimulating the production of basic fibroblast growth factor and vascular endothelial growth factor by macrophages and endothelial cells. This is an amazing property which was not demonstrated in others organic acids.

In vitro and in vivo apoptosis of PMNs induced by *P. aeruginosa* was inhibited by whole cultures of *L. plantarum* (Peral et al 2009). Our results demonstrate that *P. aeruginosa* supernatants induced PMN cytotoxicity and this induction was higher in the case of a clinically isolated strain when compared to a non-AHL-producer strain. Because of this,

AHL could be involved in the pathogenic activity of *P. aeruginosa* (Smith et al 2002; Tateda et al 2003). This cytotoxicity induced a remarkable drop in pHi, which was independent of the pHo. These results agree with observations by other authors who found that *P. aeruginosa* induced apoptosis and pHi decreased remarkably during the process (Tateda et al 2003; Usher et al 2002; Zaborina et al 2000; Trevani et al 1999). Our results also show that *L. plantarum* supernatants partially inhibited cytotoxicity produced by *P. aeruginosa*.

The PMN cytotoxic effect induced by bacterial supernatants, mixtures of supernatants and AA was correlated with the inflammatory response in the skin of mice. *P. aeruginosa* supernatants induced a higher inflammatory response, particularly supernatants containing AHL, as reported previously (Tateda et al 2003). At the same pH, SLp produced a weaker inflammatory reaction than AA. Neutralization of SLp (SLpN) diminished the inflammation. When *P. aeruginosa* and *L. plantarum* supernatants (acidic and neutralized) were mixed and inoculated, inflammations were less severe than inoculums of pure *P. aeruginosa* supernatants.

In summary, our studies show that *P. aeruginosa* supernatants promote inflammatory processes partly because they are highly cytotoxic to PMNs and perhaps to other cells involved in tissue repair. Chronic wounds are characterized by a constant influx of PMNs because of the biofilm mode of bacterial infection. PMNs release cytotoxic enzymes, free oxygen radicals and inflammatory cytokines that cause extensive additional damage in the wound tissue. Antibiotic treatment and antimicrobial activity by PMNs, the first defense line of the innate immune system, are inefficient to eliminate biofilm bacterial infection (Costerton et al 1999). This is the main reason for using antiseptics in the case of chronic and infected wounds. However, antiseptics are cytotoxic themselves and cause delay in wound healing so they are used with certain reservations (Cooper et al 1991). SLp and SLpN were less cytotoxic than AA, and inhibited cytotoxicity and inflammation induced by *P. aeruginosa* supernatants.

Previously we reported that *L. plantarum* cultures have a remarkable capacity to inhibit the synthesis of quorum-sensing signals, biofilm and elastase by *P. aeruginosa*. Currently, bacteriotherapy with *L. plantarum* for burns and chronic wounds is being carried out at our hospital (Valdez et al 2005; Ramos et al 2008; Peral et al 2009)

Further studies on the mechanisms exerted by *L. plantarum* regarding the tissue repair process are required to improve its performance.

COMPARISON OF VIRULENCE CAPACITY OF PLANKTONIC AND BIOFILM *P. AERUGINOSA* IN A MURINE MODEL BURN

We investigated the differences in the behavior of *P. aeruginosa* obtained from a human chronic infections, a clinical strain sample containing multiple genetic variants (polyclonal) (Erlich et al 2004; Head and Yu 2004), compared to a single clone of *P. aeruginosa* (monoclonal) isolated from the polyclonal sample.

The *P. aeruginosa* clinical strains (C) used in this study were obtained from a chronic burn wound of one month of evolution. The infection was resistant to several kinds of treatment, even with antibiotics. The biopsy sample was homogenized and cultured in Luria–Bertani (LB) medium (Gibco, Rockville, MD, USA) once. Aliquots were kept in glycerol 40% at -80° C until use. We assumed that this procedure allowed us to preserve the original P. aeruginosa strain mixture (polyclonal) present in the wound (Erlich et al 2004; Head and Yu 2004). For single clone isolation, one CFU was picked from the plate original sample biopsy culture in LB agar and cultured again in the same medium. The passage and isolation of the selected CFU was repeated three times. Then the CFU was cultured in LB broth for 4 h at 37 8C. This isolated strain (I) was aliquoted and kept in glycerol 40% at -80° C until use.

In order to determine the infective capacity and virulence of the clinical and isolated strains (biofilm and planktonic forms), the burned-mouse model of Stieritz and Holder, modified by Rumbaugh et al. (Rumbaugh et al 1999), was used. The protocol was approved by the Animal Care and Use Committee at Texas Tech University Health Science Center. Briefly, mice were anaesthetized, their backs were shaved, and then were placed in water at a temperature of 90 8C for 10 s to burn 15% of the body surface. Immediately following the burn, the mice were randomized into four groups and injected subcutaneously directly under the burn with *P. aeruginosa* with polyclonal and monoclonal sample, in both planktonic and biofilm forms.

To prepare the injection samples, the biofilm was isolated from planktonic cells from a 4-h culture, allowing biofilm to adhere to the wall surface of flask culture. The suspension of biofilm bacteria was performed by disruption using a vortex and adjusting the DO600 to the same DO600 of the planktonic suspension. We assumed that the number of bacteria in the biofilm suspension is the same in polyclonal and monoclonal biofilms and similar to that in the planktonic suspension. The injected groups were: (1) BPaCp group, with 200–300 CFU of planktonic *P.aeruginosa* from clinical strains; (2) BPaCb group, injected with a suspension of *P. aeruginosa* biofilmfrom clinical strains equivalent to 200–300 CFU of planktonic cells; (3) BPaIp group, injected with 200–300 CFU of planktonic *P. aeruginosa* from the isolated strain and (4) BPaIb group, injected with a suspension of *P. aeruginosa* biofilm from the isolated strain equivalent to 200–300 CFU of planktonic cells. The control group was injected with phosphate buffered saline (PBS).

Forty-eight hours after the initial infection, mice were killed with ether. In order to determine the bacterial load, the burnt skins, livers and spleens were aseptically removed and homogenized in sterile PBS by using a hand-held tissue grinder. The tissue homogenates were serially diluted in PBS and plated for growth in Mac Conkey agar. The inoculated plates were incubated at 37 8C overnight. The number of CFU was determined.

The survival pattern of the mice injected with the different samples is shown in Figure 22. Burn wound infection with planktonic cells proved to be more lethal than with the biofilm form for both the clinical and the isolated strain ($p<0.05$). There were no significant differences in survival percentage between the planktonic form of the clinical and the isolated colony or between the biofilm of the clinical and the isolated colony. Microbiological comparisons of the CFU in skin, livers and spleens of different groups revealed greater virulence for the planktonic than for the biofilm forms in the skin ($p<0.001$) with strong dissemination to livers and spleens ($p<0.0001$). No significant differences were detected between the biofilms of clinical or isolated strains. In contrast, the planktonic isolated form showed a higher number of bacteria in the organs than the planktonic clinical form ($p<0.001$) (Table 3).

Infected burn wound healed spontaneously after 30 days in the mice that survived the infection both for the clinical sample and for the colony isolated from it whether in biofilm or in planktonic form. No significant differences, was found among these groups.

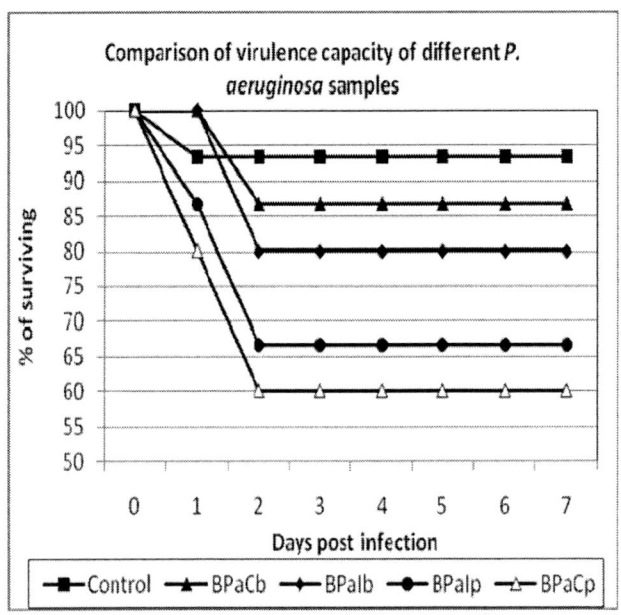

Figure 22. Survival curves of mice for burn injected with saline (control) and different samples of P. aeruginosa: PaCB, burned and infected with biofilm of *P. aeruginosa* clinical strain; PaIB, burned and infected with biofilm of *P. aeruginosa* isolated strain; PaCP, burned and infected with planktonic of *P. aeruginosa* clinical strain; PaIP, burned and infected with planktonic of *P. aeruginosa* isolated strain.

Table 3. Microbiological analyses of tissues from mice infected with different samples of *P. aeruginosa*: PaCB, burned and infected with biofilm of *P. aeruginosa* clinical strain; PaIB, burned and infected with biofilm of *P. aeruginosa* isolated strain; PaCP, burned and infected with planktonic of *P. aeruginosa* clinical strain; PaIP, burned and infected with planktonic of *P. aeruginosa* isolated strain

	CFU (skin)	CFU (spleen)	CFU (liver)
PaCB	1.32×10^5	6	4
PaIB	1.97×10^5	8	3
PaCP	1.23×10^7	1.13×10^5	1.26×10^5
PaIP	1.46×10^7	1.29×10^6	1.15×10^6

L. *PLANTARUM* EFFECT ON INFECTION IN A MURINE MODEL BURN

In order to investigate the in-vivo *L. plantarum* inhibitory effect, a burned-mouse model of *P. aeruginosa* infection was treated with whole cultures of *L. plantarum*.

The burned-mouse model of Stieritz and Holder, modified by Rumbaugh et al. (Rumbaugh et al 1999) above mentioned was used. To infection with *P aeruginosa* mice were injected subcutaneously directly under the burn with 200–300 CFU of *P. aeruginosa*/100 µL (group BPs). To treatment with L plantarum (group BPs + Lp), mice were injected in the burned and infected area, on days 3, 4, 5, 7, and 9 after the initial infection, with 100 µL of a suspension of *L. plantarum* (10^5 CFU/mL) grown in MRS broth.

On days 5, 10 and 15 after the initial infection, mice were anaesthetised with ether, blood was collected, and skin, connective tissue and panniculus carnous muscle from the burned area were processed. Blood, skin and connective tissue sections from the burned area, spleen and liver of each animal were obtained and the CFU/g of tissue was determined by plating on MacConkey agar and MRS agar.

Tissue phagocytes (TPs) were obtained from the connective and muscular tissue sections underneath the burned skin by enzymatic digestion tuissue. The adherent cells from the cell suspensions (mostly PMNs) were isolated by allowing them to adhere to the plastic surfaces of Petri dishes. TP phagocytosis of *P. aeruginosa* and *L. plantarum* was detected by indirect immunofluorescence, using specific polyclonal antibodies against each microorganism. The cells were analysed by double or triple fluorescence under an epifluorescence microscope. *P. aeruginosa* cells inside phagocytes were detected by cytoplasmic red fluorescence (Rodamine TRICT-conjugated; Jackson Immuno Research Laboratory, West Grove, PA, USA), while L. plantarum cells were detected by cytoplasmic green fluorescence (FITC-conjugated; Sigma), and the morphology of nuclei was visualized by light blue fluorescence (DAPI; Molecular Probes, Eugene, OR, USA). TP apoptosis was analyzed for the presence of DNA fragmentation by the TUNEL assay.

At day 10, treatment with *L. plantarum* (group BPs + Lp) enhanced *P. aeruginosa* phagocytosis by tissue phagocytes significantly, led to a decrease in apoptosis and a decrease in the bacterial counts in skin (Figures 23, 24 and 25).

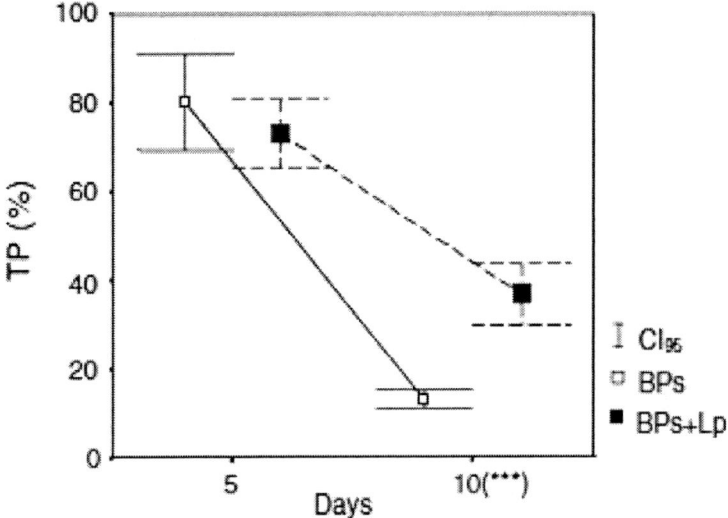

Figure 23. Tissue phagocytosis (TP) of *Pseudomonas aeruginosa* at days 5 and 10 in the burned-mouse model. Each point is the average of five experiments and each experiment was performed with cells from one mouse. BPs, burn wounds infected with *P. aeruginosa*; BPs + Lp, burn wounds infected with *P. aeruginosa* and treated with *L. plantarum*. ***$p < 0.001$; CI95, 95% CI for the mean.

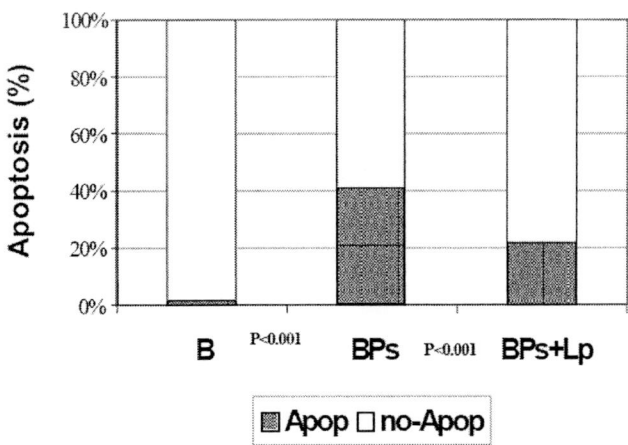

Figure 24. Apoptopsis of Tissue phagocytes (TP) at days 10 in the burned-mouse model. Treatment with *L plantrum* protected phagocytes from apoptosis induced *P. aeruginosa* .

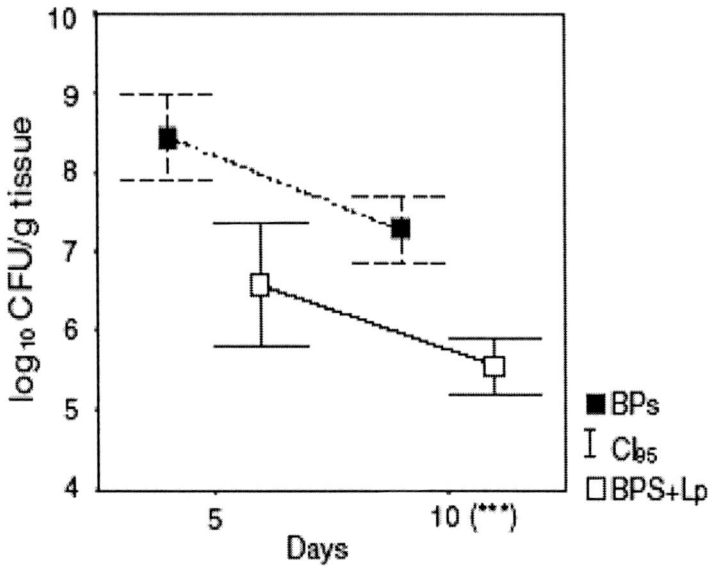

Figure 25. Comparison of *Pseudomonas aeruginosa* viable counts in tissue obtained from burn wounds at days 5 and 10 post-infection. Values represent the average of five independent experiments. BPs, burn wounds infected with *P. aeruginosa*; BPs + Lp, burn wounds infected with *P. aeruginosa* and treated with *Lactobacillus plantarum*. ***$p < 0.001$; CI95, 95% CI for the mean.

The effect of the treatment was more pronounced on day 15, with macroscopically diminished purulent exudates, a more diffuse inflammatory cell infiltrate, and fast regeneration and tissue repair.

There were not isolated on the analized samples *L. plantarum* by retrocultures.

It results shown that in this animal models of infection the ability of *P. aeruginosa* to express virulence factors and elements which promote its adherence to the damaged tissue and form biofilm, and affect the activity of PMNs neutrophil functions keeping the wounds arrested in a stage dominated by inflammatory processes, was attenuated by *L. plantarum* treatment. Bacteria in biofilms express quorum-sensing-controlled virulence factors that can kill or eliminate the activity of host immune cells. It has been shown that rhamnolipid, a leukocidal toxin produced by *P. aeruginosa*, causes rapid necrosis of PMNs in vitro. *P. aeruginosa* biofilm in chronic wounds were surrounded by host cells, but were not penetrated (Fazli et al 2009).

Chapter 5

CLINICAL STUDIES

Our decision to use *L. plantarum* in our clinical experiments was based on the fact that there are no reports indicating the possible virulence activity and the capacity to improve wound healing of *L. plantarum* in a burn experimental models (Valdez et al 2005) or in spontaneous infections (Reid et al 2002). Although lactobacilli have an excellent overall safety record among probiotics, there are a few reported cases of infection in premature neonates with severe immune deficiencies caused mainly by *Lactobacillus rhamnosus* (Boyle et al 2006).

BURNS TREATMENT

Early excision and grafting in burn wound treatment have become the norm. However, the suboptimal facilities that prevail in developing country hospitals added to insufficient funds, inadequate information, poor management and lack of personnel result in delays in wound treatment. Burn wound infections occur more frequently in countries with overcrowded burn units, fewer infection control barriers and less access to immediate wound debridement or antimicrobial therapies. Consequently, antimicrobial therapy is important in the treatment and prophylaxis of wound infections. In our unit, as in other burn care facilities, antibiotics are not routinely administered because of their cost and of the high degree of antibiotic resistance. The persistence of these infections which not resolves with surgical debridement, antiseptics and antibiotics lead us to develop an adjuvant treatment of topical application of lactobacilli that in vitro had a great antibacterial activity.

The burn wound surface is sterile immediately following injury; however, it is repopulated quickly with gram-positive organisms from the environment during the first 48 hours. More virulent gram-negative organisms replace the grampositive organisms after 5–7 days. Burns produce disruption of the mechanical integrity of the skin and generalized immune suppression that allows micro-organisms to multiply freely.

Currently, the common pathogens isolated from burn are Staphylococcus aureus, found in 75% of wounds, *Pseudomonas aeruginosa* (25%), *Streptococcus pyogenes* (20%) and various coliforms (5%). Other streptococci, anaerobic organisms and fungi can also cause infections (Edwards-Jones and Greenwood 2003). The above bacteria are producers of biofilms, so they are extremely resistant to antibiotics, antiseptics and to the host immune response.

In the case of infected or necrotic tissue, the wound bed must be prepared either using surgical debridement or by a progressive local treatment that eliminates dead tissue. Topical antimicrobial agents are used to prevent infection. Povidone iodine is used in the bath (Edwards-Jones et al 2000) and Silver sulphadiazine (SD-Ag) is the gold standard in topical burn treatment (Atiyeh et al 2007). In most burn care units, antibiotics (erythromycin or flucloxacillin), used for the prevention of wound infection, are not routinely administered to burn patients because of their cost, the high incidence of resistance and the risk of adverse drug effects (Edwards-Jones and Greenwood 2003). Alternative medical therapy such as natural medicines and old-fashioned treatments are resurfacing. Sugar, honey, carica papaya fruit and aromatherapy with aromatic oils have been effectively used as dressings and for the debriding and cleansing of infected wounds. Some of these medicines, however, may be toxic and are not recommended for use on broken skin (Cooper and Molan 1999).

On these bases, we performed an exploratory study to establish the effectiveness of bacteriotherapy with topical application of the innocuous bacteria *Lactobacillus plantarum* cultured in De Man, Rogosa and Sharpe medium to provide an alternative method for burn treatment using SD-Ag as a reference (Peral et al 2009).

Patient enrolment was conducted so as to obtain two sufficiently homogeneous groups with respect to age, gender, bacterial load, localization and type of burns (second and third degree and % of total body surface area (TBSA) as described in Table 4.

Table 4. Characteristics of the patients treated with *L. plantarum* or SD-Ag

	Patient (n)	Degree and %TBSA mean (range)			Age (range)	sites	bacteria/g tissue(range)
		II D	III E	III D			
L. plantarum	14	12(6-24)			35(18-55)	Arms, hands, legs, abdomen, thorax	$3 \times 10^7 (3 \times 10^6 - 8 \times 10^7)$
L. plantarum	12		10(2-15)		35(18-50)	Arms, leg, hands, neck	Non infected
L. plantarum	12			11(2-15)	39(22-58)	Arms, leg, hands, thorax	$9 \times 10^6 (6 \times 10^6 - 10^8)$
SD-Ag	15	16(8-22)			30(18-55)	Leg, thorax, abdomen, arms, hands, face, neck	$8 \times 10^6 (2 \times 10^6 - 6 \times 10^7)$
SD-Ag	13		9(4-12)		45(19-59)	Leg, thorax, abdomen, hands	Non infected
SD-Ag	14			8(2-15)	37(18-60)	Thorax, abdomen, hand, leg, arms	$10^7 (2 \times 10^6 - 7 \times 10^7)$

Delayed second degree (II D) and delayed third degree (III D) burns. These patients had their first surgery at an average of 5 days post burn (range 3–7 days).

Early third degree (III E) burns. These patients began treatment (underwent first excision when necessary) at an average of 1day post burn (range 0-2 days).

Before treatment and wounds showed clinical evidence of infection with microbial counts of 106–108 bacteria/g tissue, and all types of bacteria were isolated: *P. aeruginosa* (51%), *S. aureus* (38%), *S. epidermidis* (10%) andothers such as *E. cloacae*, *K. pneumoniae* and *E. faecalis* (2%). *P. aeruginosa* and *S. aureus*, the most frequently isolated bacteria, produced biofilm when they were assayed by biofilm formation in 96-well polyvinyl chloride microtiter plates stained with 0,1% crystal violet technique (O´Toole and Kolter 1998).

Each patient was allocated to receive, in random order, treatment with *L. plantarum* or SD-Ag as described below.

A whole culture of 10^5 *L. plantarum* (ATCC) 10241/ml in the log phase grown in MRS broth (Oxoid, Basingstoke, UK) at 37°C was used. A volume of the culture large enough to cover the wound (1 ml/ cm2) was spread on a gauze pad and applied to the burn, which was then routinely bandaged.

Moisture was optimum until the next day. This procedure was used once a day over a period of 10 days. New cultures were prepared every day. The wounds treated with L. plantarum were not treated with any traditional antimicrobial.

Fifteen patients with delayed second-degree burns, 13 early third-degree burns and 14 delayed third-degree burns were treated with 1% SD-Ag cream (Denver®; Farma Lab, Buenos Aires, Argentina). Patients received a daily antiseptic bath with chlorhexidine 0,05%, and then, the burn wounds received a 3-mm layer of SD-Ag cream every 24 hours for 10 days. The cream caused an exudate by wound maceration.

Clinical and bacteriological follow-up was performed. All the results are summarized in Table 5.

This preliminary trial of this treatment suggests no differences between treatment with L. plantarum and with SD-Ag in infected second-degree and non infected third-degree burns. In infected third-degree burns, treatment with *L. plantarum* would show greater efficacy.

The effect of the *L. plantarum* culture on the wounds observed in this study could be because of the fact that the cytokine pattern induced by *L. plantarum* in inflammatory cells is the opposite of those induced by pathogens

Acid pH contributes to the antimicrobial activity and also it has also been reported to contribute to the activation of the cells involved in the immune response and in tissue repair (Lardner 2001). In a previous study, we investigated the interference of *L. plantarum* with *P. aeruginosa* in controlled experiments. We determined that L. plantarum inhibited in vitro the activity of N-acyl-homoserine lactone and bacterial growth as well as the formation of biofilm and elastase by *P. aeruginosa*. The in vivo application of *L. plantarun* to burns infected with the pathogen inhibited colonization, modifying the inflammatory response, and promoting tissue repair in a burn murine model.

As this is a first approach to the subject, further studies on the treatment and prophylaxis of burn infections with *L. plantarum* will be required before establishing the *L. plantarum* treatment in the context of current treatments.

Table 5. Comparison of patients treated with *L. plantarum* or SD-Ag

	Proportion of patients (p)		Relative rates[a]
Treatment	*L. plantarum*	SD-Ag	
IID* (n° patients)	14	15	
Graft	no	no	
bacteria/ g tissue $<10^5$ $>10^5$	n=14 p=0.71 10= $6 \times 10^3 \pm 5 \times 10^3$ 4= $7 \times 10^6 \pm 5.2 \times 10^6$	n=15 p=0.73 11= $3.6 \times 10^4 \pm 2.3 \times 10^4$ 4= $3.4 \times 10^6 \pm 3 \times 10^6$	-2.73 %
Granulation tissue (range 7-13 days)	n=14 p=0.71 (10/14)	n=15 p=0.73 (11/15)	-2.73 %
Healing (range 30-50 days)	n=14 p=0.71 (10/14)	n=15 p=0.73 (11/15)	-2.73 %
IIIE[#] (n° patients)	12	13	
Granulation tissue (range 7-13 days)	n=12 p=0.75 (10/12)	n=13 p=0.84 (11/13)	-1.07%
Graft taking (range 6-10 days)	n=10 p=0.90 (9/10)	n=11 p=0.90 (10/11)	0%
Healing (range 30-50 days)	n=12 p=0.75 (9/12)	n=13 p=0.77 10/13	-2.60%
IIID** (n° patients)	12	14	
bacteria/ g tissue[a] $<10^5$ $>10^5$	n=12 10= $7 \times 10^3 \pm 4 \times 10^3$ 2= $7 \times 10^6 \pm 2 \times 10^5$	n=14 10= $2.5 \times 10^3 \pm 2 \times 10^3$ 4= $5 \times 10^6 \pm 4.4 \times 10^6$	+16.90%
Granulation tissue (range 7-13 days)	n=12 p=0.83 (10/12)	n=14 p=0.71 (10/14)	+16.90%
Graft taking (range 6-10 days)	n=10 p=0.90 (9/10)	n=10 p=0.90 (9/10)	0%
Healing (range 30-50 days)	n=12 p=0.75 (9/12)	n=14 p=0.64 (9/14)	+17.19%

* Delayed second degree and **delayed third degree burns. [#] Early third degree burns.
[a] Relative rates = $\frac{p\ L.\ plantarum - p\ SD\text{-}Ag}{p\ L\ plantarum} \times 100$.

ULCER TREATMENT

The problem of chronic wounds is large in scope as well as in psychosocioeconomic cost. There are many causes of chronic wounds, with diabetes, pressure ulcers, and venous stasis as the 3 most common causes. In developed countries, it has been estimated that 1 to 2% of the population will experience a chronic wound during their lifetime (Gottrup 2004). Chronic wounds are, by definition, wounds that remain in a chronic inflammatory state and therefore fail to follow the normal patterns of the healing process. The normal function of inflammation in an acute wound is to prepare the wound bed for healing by removing necrotic tissue, debris, and bacterial contaminates as well as recruiting and activating macrophages and fibroblasts. Under normal conditions, inflammation is a self-limiting process. In an acute wound, activated neutrophils are virtually nonexistent after the first 72 hours, whereas in a chronic wound, neutrophils are present throughout the healing process (Menke et al 2007). Polymorphonuclear neutrophil (PMN) leukocytes are probably the most significant components of the host defence mounted against biofilm-forming bacteria, and their secretory and phagocytic arsenal often fails to eliminate bacteria in biofilm. Chronic wounds are rarely, if ever, sterile and achieving wound sterility is often an unrealistic and non-essential goal in wound care. Bacteria infecting chronic wounds are producers of biofilm and so are extremely resistant both to antibiotics and to the host immune response (Costerton et al 2003).The possible reasons for the persistent presence of neutrophils include continued recruitment and activation by bacterial overgrowth and leukocyte trapping. Microbial products induce the secretion of inflammatory mediators. One of the most important is interleukin (IL)-8, a potent chemoattractant of neutrophils.

The result of continued up-regulation of the inflammatory cascade leads to a markedly abnormal inflammatory profile for chronic wounds. The large number of activated neutrophils leads to excessive amounts of degradative matrix metallic proteinases (MMPs), because the MMPs are not balanced by an equal amount of tissue inhibitors of matrix metalloproteinases. Neutrophil-derived elastase is an abundant primary granule serine proteinase with the ability to degrade a wide range of connective tissue macromolecules. In addition, the collagenase MMP-8 is a major constituent of neutrophil secondary granules. The collagenases are critical enzymes in the breakdown of the extracellular matrix because they are unique in their ability to initiate degradation of collagen. There is growing evidence suggesting that these enzymes may also play a role in the pathophysiology of chronic wounds by not only degrading the synthetic products of fibroblasts and peptide growth factors such as transforming growth factor b and PDGF but also important wound components such as fibronectin, a-1 antiprotease, and a-2 macroglobin. In this way, infection and inflammation contribute to the chronicity of the wound. In fact, neutrophils contribute to tissue lesion and debris production, facilitating bacterial proliferation (Bjarnsholt et al 2008). Chronic wound infections often do not respond to traditional antimicrobial therapies (Percival and Bowler 2004), Antibiotic and antiseptics by themselves are insufficient to manage infections in chronic wounds. For instance, debridement, pressure relief, and moisture-retentive dressings can reduce the likelihood of infection. (Landis 2008).

Because excessive inflammation is the ultimate cause of the poor healing found in chronic wounds, most treatments are aimed at reducing inflammation. Surgical debridement and wound care methods are aimed at decreasing the necrotic tissue and protease burden, thus providing a virtual resetting of the wound back into the acute healing phase. If the inflammation level is subsequently kept low, the wound is then able to progress forward and begin to heal. Some methods for altering the inflammatory cascade such as the using of exogenous cytokines and growth factors to shift the degradative may disrupt the inflammatory cycle and allow normal progression of the wound healing process (Menke et al 2007).

Financial and operative limitations in our hospitals have led us to investigate alternative therapies. One of them is the bacteriotherapy with *Lactobacillus plantarum* based in the experimental (Valdez et al 2005) and clinical (Peral et al 2009) results detailed above.

In this study, we evaluate the efficacy of Bacteriotherapy with *L. plantarum* culture on the chronic infected leg ulcers of diabetic and non-diabetic patients and to observe its effects on apoptosis, necrosis and IL-8 production by the cells from the ulcer bed (CUB). In addition, IL-8 production was analyzed in peripheral blood PMN (PBPMN) from patients with ulcers and compared with that in PBPMN from normal and diabetic subjects without ulcers under basal conditions, and after infection with *P. aeruginosa* either with or without preincubation with *L. plantarum*.

The study comprised male and female individuals aged 40–70 years of age. Thirty-four patients from the plastic surgery and burns unit of the 'Zenon Santillan' Hospital with a chronic venous ulcer were included in a prospective uncontrolled study employing a local *L. plantarum* treatment.

Fourteen patients suffered from moderately controlled type 2 diabetes mellitus (glycaemia level: 1.50 ± 0.30 mg/mL, HbA1C: $7.8 \pm 2.1\%$) and the 20 remaining patients were non-diabetic. Inclusion criteria included the presence of one venous ulcer confirmed by venous duplex ultrasound, with a surface of 25–60 cm2, a bacterial load at a level $>10^5$ microorganisms per gram of tissue, which is generally accepted to justify a diagnosis of infection and is an important factor in delayed healing in chronic wounds (Landis 2008) and no signs of healing in the past 3 months, despite conventional medical treatment. Inclusion criteria comprised patients who had malignancy, autoimmune disease, an inclination to bleed or bleeding disease, and serious systemic infection.

Ten diabetic patients with similar glycaemia levels without lesions and 14 healthy subjects with normal glycaemia levels attending the 'Angel C. Padilla' Hospital were included as PBPMN donors. Both hospitals are located in the city of San Miguel de Tucuman, Argentina. All patients were informed about the aims of the study and provided their consent. The study was approved by the Hospital's Ethics Committee.

To treatment with *L. plantarum* wounds were cleaned, irrigated with saline and treated with topical applications of a whole culture of 10^5 *L. plantarum* ATCC 10241/mL in log phase, which was previously grown in De Man, Rogosa and Sharpe (MRS) broth for 5–6 h at 37°C. The culture was spread on a gauze pad and applied to the lesion, which was then covered with occlusive dressing. The culture was applied once-daily over a period of 10 days.

Tolerable discomfort such as a burning sensation was observed after the first application of *L. plantarum*. The lesions were clinically monitored and evaluated weekly by the plastic surgeon

To obtain lesion biopsy samples a 4 mm^3 of tissue was taken. One part of the biopsy was enzymatically digested to obtain CUB. Other part of the biopsy was processed for microbial evaluation by routine techniques. Also, Forty-eight hours after the 10-day treatment period, wound and blood samples were taken and incubated in MRS broth in an attempt to recover *L. plantarum*.

To obtain PBPMN heparinized venous blood samples were collected from all individuals. In the case of patients with ulcers, peripheral blood samples were obtained before performing bacteriotherapy.

Neutrophils were isolated by dextran T-500 and Ficoll- Hypaque gradient centrifugation as indicated above. The cells were suspended at 10^6 PBPMN/mL in RPMI 1640-HEPES Medium supplemented with foetal bovine serum 10% v/v. Cell cultures were infected for 1 h with 2×10^7 *P. aeruginosa* standard clinical isolate. Cells were fixed and IL-8 was determined.

For the study of L. plantarum interference with *P. aeruginosa*, PBPMN cultures (10^6/mL) were preincubated for 1 h at 37°C with 2×10^5 *L. plantarum* and subsequently infected with 2×10^7 *P. aeruginosa* for 1 h at 37°C.

In all cases cells CUB and PBPMN were fixed stained with H-E, Intracellular IL-8 was determined by an immunoperoxidase, apoptosis by TUNEL.

As shown in Table 6 the application of *L. plantarum* cultures to wounds induced, in approximately 8 days (range, 6– 10 days), a coverage of granulation tissue of more than 90% in 50% and 55% of the diabetic and non-diabetic patients, respectively. After 30 days of treatment, a reduction of more than 90% of the lesion area was observed in 43% and 50% of the diabetics and non-diabetic patients, respectively.

Table 6. Effect of *L. plantarum* treatment on granulation tissue and area of ulcers of ulcers of diabetics and non diabetics patients. (p=proportion of patients)

Groups	Diabetics	Non Diabetics
N° patients	14	20
Granulation tissue % of area coverage (media/range) Before/after 10 days treatment	n= 7 p=0.50 Before: 8% (0-12%) After: 94% (90-100%)	n=11 p=0.55 Before: 6% (0-10%) After: 96% (92-100%)
	n= 7 p=0.50 Before: 7% (0-10%) After: 39% (30-60%)	n=9 p=0.45 Before: 8% (0-10%) After: 43% (34-58%)
Ulcer Area reduction percentage 30 days of treatment Percentage (range) Size (cm^2) median/media/(range) before/after treatment	n=6 p=0.43 93% (80-100%) *Before: 26/32/(25-50)* *After: 1,2/2,2/(0-6,2)*	n=10 p=0.50 92% (80-100) *Before: 38/39/(25-54)* *After: 4/5/(0-10)*
	n=8 p=0.57 52% (range 35-70%) *Before: 46/46/(26-60)* *After: 26/24/(15-30)*	n=10 p=0.50 51% (range37-75) *Before: 48/47,6/(27-60)* *After: 25/25/(11-37)*

As shown in Figure 26A, the difference between the values at day 0 and at day 30 of wound closure area in both groups was statistically significant (p <0.001).

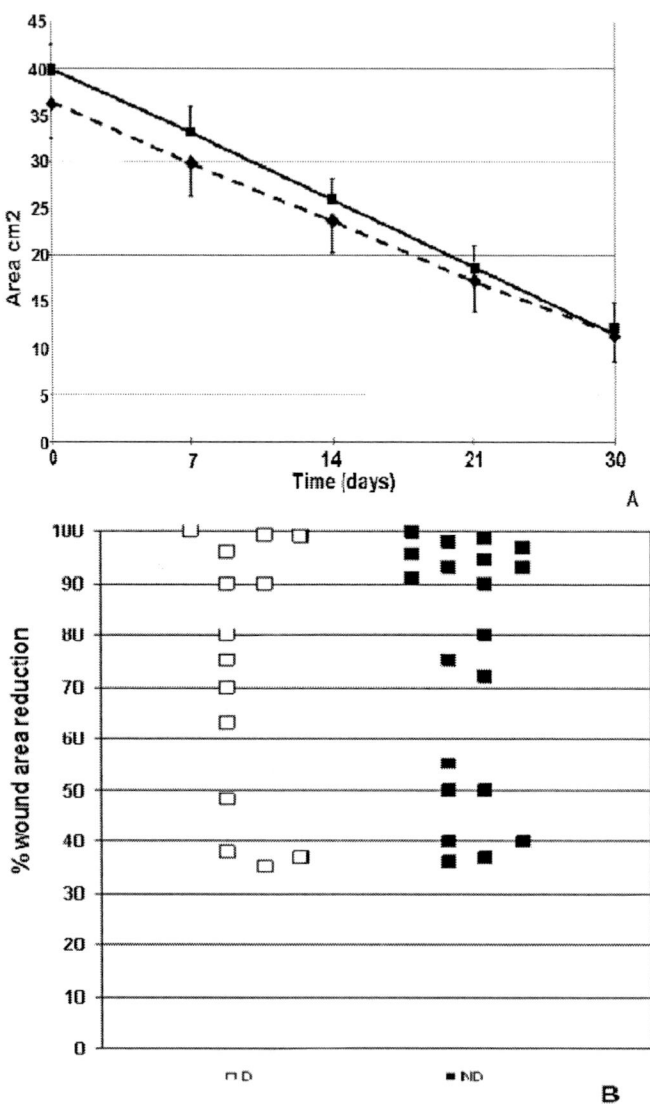

Figure 26. (A) Ulcer area size and time of treatment in diabetic (filled line) and non-diabetic (dotted line) patients. Each point represents the mean ± SD wound area (diabetics n = 14; non-diabetics n = 20). (B) Each symbol represents the percentage of wound reduction area in each individual patient on day 30 after the start of the treatment. Open squares, diabetics (D); filled squares, non-diabetics (ND).

The degrees of the reduction of the wound areas 30 days after the start of treatment in each wound study of the diabetic and non-diabetic patients is shown in Figure 26B. This indicates a continuous healing process. No significant differences were observed between the diabetic and non-diabetic groups.

There was a remarkable decrease in bacterial load with treatment. Before treatment with *L. plantarum*, wound microbial counts were 5×10^5 to 10^8 bacteria/g tissue. The bacteria isolated included *Staphylococcus aureus* (45%), *P. aeruginosa* (35%) and *Staphylococcus epidermidis* (15%), as well as others such as *Enterobacter cloacae*, *Klebsiella pneumoniae* and *Enterococcus faecalis* (5%). All these bacteria were biofilm producers when assayed for biofilm formation in 96-well polyvinyl chloride microtitre dishes stained with 0.1% crystal violet. The number of CFU was lower than 10^5 after 5 days of treatment and dropped further (to $<10^3$) after 10 days (p <0.001). No significant differences were observed between diabetic and non-diabetic patients (p=0.97). Forty-eight hours after the end of the treatment with L. plantarum, this bacterial species was not recovered from either peripheral blood or wound samples.

The IL-8 changes in CUB with *L. plantarum* treatment in samples were taken on days 0, 5 and 10 from the deep bed tissue. Most cells on days 0, 5 and 10 were PMN (Figure 27).

The percentage of CUB positive for IL-8 in diabetics and non-diabetics was similar before treatment (day 0) (Figure 28), which is in agreement with a previous study (Galkowska et al 2006).

IL-8 values increased after 5 days of treatment (p <0.001) followed by a decrease after 10 days compared to 5 days of treatment (p <0.05). These variations were independent of the diabetic or non-diabetic condition of the patients. IL-8 was expressed mainly by PMN.

As shown in Figure 29, IL-8 in PBPMN of healthy individuals (N) had basal values significantly lower than basal values of all other groups, diabetic patients without ulcers (D), diabetic patients with ulcers (DU) and non-diabetic patients with ulcers (NDU) (p <0.001). After *P. aeruginosa* infection, the percentage of IL-8 positive PBPMN was increased significantly in groups N, D and DU compared to basal values (p <0.001) and, to a lesser extent, in the NDU group (p <0.01). There were no significant differences among D, DU and NDU groups. Preincubation of PBPMN with *L. plantarum* prior to infection with *P. aeruginosa* inhibited the number of IL-8 positive cells compared to cells that were not pre-incubated only in the DU and NDU groups (p <0.001). No significant differences were observed between D and DU patients.

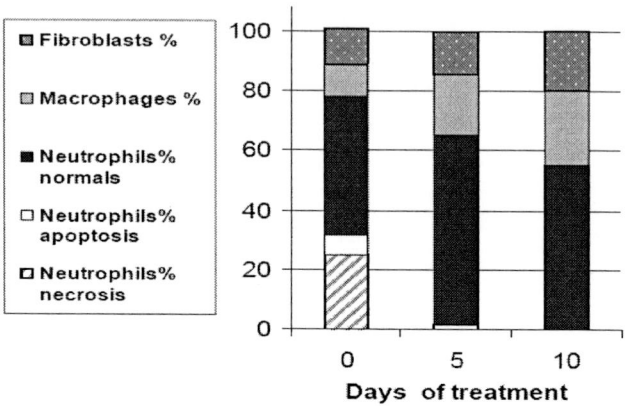

Figure 27. Variation in cell subpopulation proportions in CBU with the treatment. Data are presented as the means of 20 values from ulcers of non diabetics plus 14 values from ulcers of diabetic patients because no significant differences were observed when both groups were compared. Standard deviations are not shown.

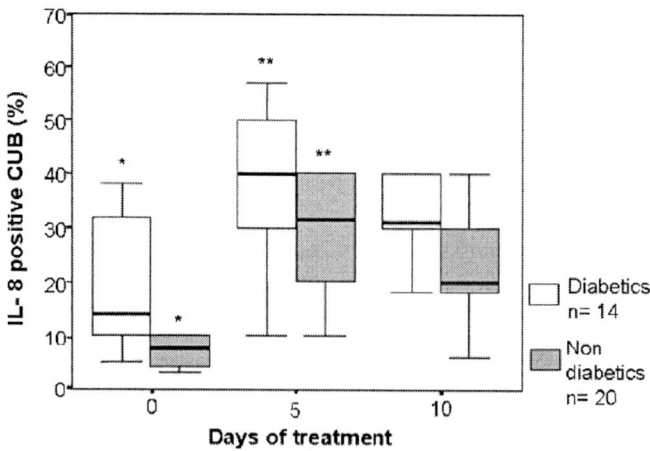

Figure 28. Percentage of interleukin (IL)-8-positive cells from the ulcer bed from non-diabetics (filled squares) and diabetics (open squares) with ulcers as a function of time. Boxes refer to the 25th (bottom) and 75th (top) percentiles, and the median is the horizontal line inside; fences refer to maximum and minimum values and open circles represent outliers. *Significantly lower compared to days 5 and 10 (p <0.001) within the same group. **Significantly higher compared to day 10 (p <0.05) within the same group.

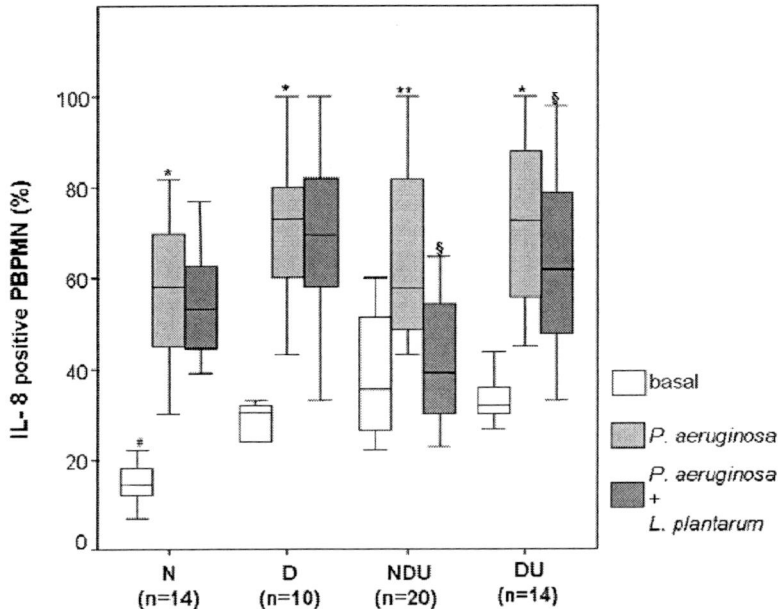

Figure 29. Percentage of positive peripheral blood polymorphonuclearneutrophils (PBPMN) for interleukin (IL)-8 production from differentgroups of individuals: normal (N), diabetics without ulcers (D), nondiabetics with ulcers (NDU), diabetics with ulcers (DU). The assaywas carried out with PBPMN under basal conditions (open squares) and with PBPMN infected with Pseudomonas aeruginosa preincubated either with (dark grey squares) or without Lactobacillus plantarum (light grey squares). Boxes refer to the 25th (bottom) and 75th (top) percentiles, and the median is the horizontal line inside; fences refer to maximum and minimum values and open circles represent outliers. Significantly lower than the other basal values: #p <0.001. Significantly higher than the basal values within the same group: *p <0.001; **p <0.01. Significantly lower than P. aeruginosa infected alone within the same group: §p <0.001.

The application of *L. plantarum* caused a debridement and granulation of wounds, a decrease both in the bacterial load and in the area of the lesion in a significant percentage of diabetic and non-diabetic patients, with no significant differences between the two groups.

Among the more important finding we observed the effects of Bacteriotherapy on the modification of the inflammatory response and on infection in chronic venous ulcer leg wounds in moderately controlled type 2 diabetes mellitus diabetic and non-diabetic patients. The wounds showed polymicrobial infections with biofilm-producing bacteria.

Lactobacillus plantarum treatment decreased apoptosis and necrosis of neutrophils and promoted the appearance of fresh PMN and other cells (i.e. fibroblasts and endothelial) in the ulcer bed. These cells were viable and exhibited changes in IL-8 production.

The greater percentage of basal IL-8 in PBPMN found in diabetics and patients with ulcers with respect to normal subjects would predispose the former to inflammation.

We found no significant differences in the behavior of the CUB and PBPMN in diabetics and non-diabetics, most likely because the former did not have advanced macro- and/or microvascular complications.

The *L. plantarum* inhibition of IL-8 induced by *P. aeruginosa* in PBPMN is in accordance with the effects observed in human leukocytes when stimulated in vitro with *P. aeruginosa* or *L. plantarum*. Bacteria may have intra- and interspecies communication for synergistic or antagonist actions (von Bodman et al 2008), as has been demonstrated in habitats such as the gut and vagina (Reid et al 2002) and the lung in cystic fibrosis (Harrison 2007).

We hypothesize that the *L. plantarum* treatment antagonize the patogens activity mainly *P. aeruginosa* and *S. aureus* the most important cualitative and cuantitative bacteria in this polymicrobial infections with biofilm-producing bacteria.

L. plantarum in vitro and in vivo activity against *P. aeruginosa* was determined in previous experimental and clinical studies mentioned above. The Lactobacillus activity against *S. aureus* was demonstrated in experimental wound infections (Gan et al 2002).

The interaction with immune inflammatory cell host with different bacteria such as *P. aeruginosa*, *S. aureus*, added to disruption of biofilm, inhibition of autoinducers and virulence factor, Lactobacillus induce different citokine patterns.

L. plantarum treatment regulates IL-8 levels induced by *P. aeruginosa* and modulates the entry and activity of the newly-arrived PBPMN to the ulcer bed. Also this interference decrease in apoptosis and necrosis and a more effective phagocytosis would diminish the number of bacteria, the debris and the inflammatory response. Consequently, the number of PMN would decrease and would promote an increase in macrophages and fibroblasts, as well as tissue repair processes.

The interference with pathogens quorum sensing signals, virulence factors, biofilm could modulate the host immune system and may be a promising strategy to manage chronic bacterial infections.

Chapter 6

CONCLUSIONS

In these studies we connected the science of biofilm, the bacterial interference, the chronic inflammatory response, and the chronic wound to support with scientific bases the application of bacteriotherapy as an alternative way of combating infections.

We observed that:

1. *L. plantarum* supernatants have a remarkable capacity to inhibit the synthesis and disrupt biofilm of *P. aeruginosa* (Figure 2 and 3).
2. *L. plantarum* supernatants can prolong the latency phase (Figure 4).
3. *L. plantarum* supernatants inhibit *P. aeruginosa* adhesion (Figure 5).
4. *L. plantarum* supernatants can inhibit the synthesis of *P. aeruginosa* quorum sensing signals (acyl-homoserine lactones) (Table 1; Figures 13 and 14).
5. *L. plantarum* supernatants can inhibit *P. aeruginosa* virulence factors (elastase) (Figure 18).
6. *L. plantarum* supernatants contain type 2 autoinducers. These molecules would realize interspecies communication with *P. aeruginosa* and modified it phenotypic expression. Or perhaps could realize a competitive inhibition by lasR and RhlR (receptors of HSLs) occupation (Figure 16).
7. *L. plantarum* supernatants have a cytotoxic effect over planktonic and biofilm cells of *P. aeruginosa* (Figures 8 and 9).

8. The pHi and viability of PMNs incubated with mixture of supernatants (SPaC-SLp, SPaC-SLpN, S129b-SLp and S129b-SLpN) were higher than pHi and viability of PMN incubated with SPaC and SP129b alone (Figure 19).
9. Incubation of PMN with acetic acid produced a higher intracellular acidocis and lesser viability than SLp (Figure 20).
10. SLp apoptosis/necrosis values were lower than acetic acid after every incubation time and at each pHo (Figure 21).
11. A lower inflammatory response (*in vivo*) was observed with a mixture of supernatants (SPaC-SLp, SPaC-SLpN, S129b-SLp and S129b-SLpN) compared to inoculation of pure *P. aeruginosa* supernatants (SPaC and S129) (Table 2).
12. The infections with *P. aeruginosa* planktonic bacteria tended to become systemic and more deadly than *P. aeruginosa* biofilm infections. All infected wounds required the same healing period (30 days). These findings were independent of the origin of the bacteria (clinical or colony isolated strain) (Figure 22; Table 3).
13. LPS was detected neither SLp nor MRS.
14. In *P. aeruginosa* infected burns, treatment with *L. plantarum* enhanced *P. aeruginosa* phagocytosis by tissue phagocytes significantly, led to a decrease in apoptosis and decrease in the bacterial load in skin (Figures 23, 24 and 25).
15. Topical application of *L. plantarum* culture in burns would be a valid alternative to other treatments like Ag-Sulfadiazine in early and delayed third-degree burns as well as in delayed second-degree burns (Table 5).
16. Topical application of *L. plantarum* culture to ulcers of nondiabetic and controlled diabetic patients induced debridement, granulation tissue formation and total healing after 30 days in 43% diabetics and in 50% non-diabetics. No significant differences between the groups were observed (Table 6; Figure 26).
17. The cells from ulcer beds collected after treatment with *L. plantarum* showed a decrease in the percentage of PMN, apoptotic and necrotic cells and produce a regulation of IL-8 production (Figures 27 and 28).
18. IL-8 production by isolated neutrophils from these patients was compared with that in diabetics without ulcers, as well as normal subjects under basal conditions, and after infection of PMN cells with *P. aeruginosa* pre-incubated either with or without *L. plantarum*. The basal values in diabetic and ulcer patients were higher than normal

and were increased by *P. aeruginosa* infection in normal, diabetics and non-diabetics with ulcers. Pre-incubation with *L. plantarum* decreased IL-8 production in patients with ulcers non-diabetic and diabetic (Figure 29).

On the basis of these findings, we hypothesize that the antipathogenic and biofilm-disrupting activity of *L. plantarum* treatment regulates IL-8 levels and modulates the entry and activity of the newly-arrived PBPMN to the ulcer bed. A decrease in apoptosis and necrosis and a more effective phagocytosis would decrease the number of bacteria, the debris and the inflammatory response. Consequently, the number of PMN would decrease and would promote an increase in macrophages and fibroblasts, as well as tissue repair processes.

The importance of this treatment lies in the easy access to and application of *L. plantarum*, its efficacy, innocuousness and low cost. In Argentinian hospitals, treatment is free and low budgets make costly treatments impossible. Because the present study comprises a first approach to the subject, with the limitation of the absence of comparator groups and the small sample size, further studies investigating the treatment of chronic infected wounds with *L. plantarum* are warranted.

ACKNOWLEDGMENTS

This work was supported by grant BID 1728 OC-AR PICT 2006 N° 1458 from Agencia Nacional de Promoción Científica y Tecnológica, Argentina and Consejo de Investigación de la Universidad Nacional de Tucumán Programa CIUNT N° 26/D453.

REFERENCES

Alakomi HL, Skytta E, Saarela M, Mattila-Sandholm T, Latva-Kala K, Helander IM. 2000. Lactic acid permeabilizes gram-negative bacteria by disrupting the outer membrane. *Appl. Environ. Microbiol* 66:2001–2005.

Algirdas JJ, Franklin MJ, Berglund D, Sasaki M, Lord CI, Bleazard JB, Duffy JE, Beyenal H, Lewandowski Z. 2003. Compromised Host Defense on Pseudomonas aeruginosa Biofilms: Characterization of Neutrophil and Biofilm Interactions. *J. Immunol.* 171: 4329–4339.

Anukam KC, Hayes K, Summers K, Reid G. 2009. Probiotic Lactobacillus rhamnosus GR-1 and Lactobacillus reuteri RC-14 may help downregulate TNF-Alpha, IL-6, IL-8, IL-10 and IL-12 (p70) in the neurogenic bladder of spinal cord injured patient with urinary tract infections: A Two-Case Study. Adv Urol 680363. Published online.

Ashahara T, Nomoto K, Watanuki M, Yokokura T. 2001. Antimicrobial activity of intraurethrally administered probiotic Lactobacillus casei in a murine model of Escherichia coli urinary tract infection. *Antimicrob Agents Chemother* 45:1751–60.

Atiyeh BS, Costagliola M, Hayek SN, Dibo SA. 2007. Effect of silver on burn wound infection control and healing: review of the literature. *Burns* 33:139–48.

Bassnett S, Reinisch L, Beebe DC. 1990. Intracellular pH measurement using single excitation-dual emission fluorescence ratios. *Am J Physiol* 258:C171–8.

Bassler B, Greenberg EP, Stevens AM. 1997. Cross-Species Induction of Luminiscence in the Quorum Sensing Bacterium *Vibrio harveyi*. *Journal of Bacteriology* 179(12):4043-4045.

Ben Harder. 2002. Germs That Do a Body Good. *Sci News* 161-72.

Bjarnsholt T, Kirketerp-Moller K, Østrup Jensen P, Madsen KG, Phipps R, Krogfelt K, et al. 2008. Why chronic wounds will not heal: a novel hypothesis. *Wound Repair Regen* 16:2–10.

Bjarnsholt T, Jensen PØ, Burmølle M, Hentzer M, Haagensen J.A.,Hougen HP, Calum H, Madsen KG, Moser C, Molin S, Høiby N and Givskov M. 2005. Pseudomonas aeruginosa tolerance to tobramycin, hydrogen peroxide and polymorphonuclear leukocytes is quorum-sensing dependent. *Microbiology* 151: 373–383.

Bloksma N, de Heer E, van Dijk H, Willers JM. 1979 Adjuvanticity of lactobacilli. Differential effects of viable and killed bacteria. *Clin Exp Immunol* 37:367–375.

Boyle RJ, Robins-Browne RM, Tang MLK. 2006. Probiotic use in clinical practice: what are the risks? *Am J Clin Nutr* 83:1256–64.

Brint JM, Ohman DE. 1995. Synthesis of multiple exoproducts in Pseudomonas aeruginosa is under control of RhlR-RhlI, another set of regulators in strain PAO1 with homology to the autoinducer-responsive LuxR-LuxI family. *J Bacteriol* 177: 7155-7163.

Britigan BE, Roeder TL, Rasmussen GT, Shasby DM, McCormick ML, Cox CD. 1992. Interaction of the Pseudomonas aeruginosa secretory product pyocyanin and pyochelin generates hydroxyl radicals and causes synergistic damage to endothelial cells. *J. Clin. Invest* 90: 2187-2196.

Bron PA, Grangette C, Mercenier A, de Vos WM, Kleerebezem M. 2004. Identification of *Lactobacillus plantarum* genes that are induced in the gastrointestinal tract of mice. *J. Bacteriol* 186:5721–5729.

Chhabra SR, Harty C, Hooi DSW, Daykin M, Williams P, Telford G, Pritchard DI, Bycroft BW. 2003. Synthetic Analogues of the Bacterial Signal (Quorum Sensing) Molecule N-(3-Oxododecanoyl)-L-homoserine Lactone as Immune Modulators. *J. Med. Chem 46:* 97-104.

Chugani SA, Whiteley M, Lee KM, Argenio DD, Manoil C, Greenberg P. 2001. QscR, a modulator of quorum sensing signal synthesis and virulence in Pseudomonas aeruginosa. *Proc. Natl Acad. Sci. USA* 98:2752-2757.

Clavel T, Haller D. 2007. Molecular interactions between bacteria, the epithelium, and the mucosal immune system in the intestinal tract: Implications for chronic inflammation. *Curr Issues Intest Microbiol* 8:25–439.

Cooper ML, Laxer JA, Hansbrough JF. 1991. The cytotoxic effects of commonly used topical antimicrobial agents on human fibroblasts and keratinocytes. *J Trauma* 31(6):775–84.

Cooper RA, Molan PC. 1999. The use of honey as antiseptic in managing Pseudomonas infection. *J Wound Care* 8:161–4.

Costerton JW, Stewart PS, Greenberg EP. 1999. Bacterial biofilms: a common cause of persistent infections. *Science* 284:1318–22.

Costerton JW, Lewandowski Z, DeBeer D, Caldwell D, Korber D, James G. 1994. Biofilms, the customized microniche. *J. Bacteriol.* 176: 2137-2142.

Costerton W, Veeh R, Shirtliff M, Pasmore M, Post C, Erlich G. 2003. The application of biofilm science to the study and control of chronic bacterial infections. *J Clin Invest* 112:1466–77.

Costerton, JW, Stewart PS, Greenberg EP. 1999. Bacterial biofilms: A common cause of persistent infections. *Science* 284:1318-1322.

Cox, CD. 1986. Role of pyocyanin in the acquisition of iron from transferrin. *Infect Immun* 52: 263-270.

Cross ML, Ganner A, Teilab D, Fray LM. 2004. Patterns of cytokine induction by gram-positive and gram-negative probiotic bacteria. *FEMS Immunol and Medical Microbiol* 42: 173–180.

Crowther JR. 2001. The ELISA guidebook. Methods in Molecular Biology, vol 149 Totowa, N.J.: Humana Press, Inc

Darouiche RO, Hull RA. 2000 Bacterial interference for prevention of urinary tract infection: an overview. *The Journal of Spinal Cord Medicine* 23: 136–141.

Darouiche RO, Donovan WH, Del Terzo M, Thornby JI, Rudy DC, Hull RA. 2001. Pilot trial of bacterial interference for preventing urinary tract infection. *Urology* 58:339-44.

Davies DG, Chakrabarty AM, Geesey GG. 1993. Exopolysaccharide production in biofilms: substratum activation of alginate gene expression by Pseudomonas aeruginosa. Appl. Environ. *Microbiol* 59: 1181-1186.

Davies DG, Parsek MR, Pearson JP, Iglewski BH, Costerton JW, Greenberg EP. 1998. The involvement of cell-to-cell signals in the development of a bacterial biofilm. *Science* 280: 295-298.

Dechaeux D, Toussaint B, Richard M, Brochier G, Croize J, Attree I. 2000. Pseudomonas aeruginosa cystic fibrosis isolates induce rapid, type III secretion-dependent, but ExoUindependent, oncosis of macrophages and polymorphonuclear neutrophils. *Infect Immun* 68: 2916–2924.

Detmar M, Brown LF, Berse B, et al. 1997. Hypoxia regulates the expression of vascular permeability factor/vascular endothelial growth factor (VPF/VEGF) and its receptors in human skin. J Invest Dermatol108:263-8.

De Keersmaecker SCJ, Verhoeven TLA, Desair J, Marchal K, J. Vanderleyden J, Nagy I. 2006. Strong antimicrobial activity of *Lactobacillus rhamnosus* GG against *Salmonella typhimurium* is due to accumulation of lactic acid. *FEMS Microbiol. Lett.* 259:89–96.

Diegelmann RF, Evans MC. 2004 Wound healing: an overview of acute, fibrotic and delayed healing. *Frontiers in Bioscience* 9: 283-289.

Duan K, Dammel C, Stein J, Rabin H, Surette MG. 2003. Modulation of *Pseudomonas aeruginosa* gene expression by host microflora through interspecies communication. *Molecular Microbiology* 50(5): 1477–1491.

Durant JA, Corrier DE, Ricke SC. 2000. Short-chain volatile fatty acids modulate the expression of the *hilA* and *invF* genes of *Salmonella* Typhimurium. *J. Food Prot.* 63:573–578.

Edwards-Jones V, Greenwood JE. 2003. What's new in burn microbiology? James Laing Memorial Prize Essay 2000. *Burns* 29:15–24.

Edwards-Jones V, Dawson MM, Childs CA. 2000. Survey into TSS in UK burns units. *Burns* 26:323–33.

Eijsink VGH, Axelsson L, Diep DB, Havarstein LS, Holo H, Nes IF. 2002. Production of class II bacteriocins by lactic acid bacteria; an example of biological warfare and communication. Antonie van Leeuwenhoek 81:639–654.

Erlich G, Hu FZ, Lin Q, Costerton WC, Post JC. 2004. Intelligent implant to battle biofilms. *ASM News* 3:127–33.

Fazli M, Bjarnsholt T, Kirketerp-Møller K, Jørgensen B, Andersen AS, Krogfelt KA, Givskov M, and Tolker-Nielsen T. 2009. Nonrandom Distribution of Pseudomonas aeruginosa and Staphylococcus aureus in Chronic Wound. *J Clin Microbiol* 47: 4084–4089.

Federle MJ, Blassler BL. 2003. Interspecies communication in bacteria. *J Clin Invest* 112: 1291-1299.

Fonder MA, Lazarus GS, Cowan DA, Aronson-Cook B, Kohli AR, Mamelak AJ. 2008. Treating the chronic wound: a practical approach to the care of nonhealing wounds and wound care dressings. *J Am Acad Dermatol* 58:185–206.

Fuqua WC, Greenberg EP. 1998. Self perception in bacteria: quorum sensing with acylated homoserine lactones. *Curr. Opin. Microbiol* 1: 183-189.

Fuqua WC, Winans SC, Greenberg EP. 1996. Census and consensus in bacterial ecosystems: the LuxR-LuxI family of quorum sensing transcriptional regulators. *Annu Rev Microbiol* 50: 727-751.

Galkowska H, Wojewodzka U, Olszewski WL. 2006. Chemokines, cytokines, and growth factors in keratinocytes and dermal endothelial cells in the margin of chronic diabetic foot ulcers. *Wound Repair Regen* 14: 558–565.

Gambello M, Iglewski B. 1991. Cloning and characterization of the Pseudomonas aeruginosa lasR gene, a transcriptionalactivator of elastase expression. *J Bacteriol* 173: 3000–3009

Gan BS, Kim J, Reid G, Cadieux P, Howard JC. 2002. Lactobacillus fermentum RC-14 inhibits Staphylococcus aureus infection of surgical implants in rats. *J Infect Dis* 185: 1369–1372.

Gan BS, Kim J, Reid G, Cadieux P, Howard JC. 2002. Lactobacillus fermentum RC-14 inhibits Staphylococcus aureus infection of surgical implants in rats. *J Infect Dis* 185: 1369–1372.

Ganin H, Tang X, Meijler MM. 2009. Inhibition of Pseudomonas aeruginosa quorum sensing by AI-2 analogs. *Bioorg. Med. Chem. Lett.* In press.

Goldberg JB, Hancock REW, Parales RE, Loper J, Cornelis P. 2008. Pseudomonas. *J of Bacteriol* 190:2649–2662.

Gottrup F. 2004. A specialized wound-healing center concept: importance of a multidisciplinary department structure and surgical treatment facilities in the treatment of chronic wounds. *Am J Surg* 187:38S–43S.

Haller D, Blum S, Bode C, Hammes WP, Schiffrin EJ. 2000. Activation of Human Peripheral Blood Mononuclear Cells by Nonpathogenic Bacteria In Vitro: Evidence of NK Cells as Primary Targets. *Infect Immun* 68: 752–759.

Harrison F. 2007. Microbial ecology of the cystic fibrosis lung. Microbiology 153: 917–992.

Hasegawa M, Yang K, Hashimoto M, Park JH, Kim YG, Fujimoto Y, Nuñez G, Fukase K, Inohara N. 2006 Differential release and distribution of Nod1 and Nod2 immunostimulatory molecules among bacterial species and environments. *J Biol Chem* 281:29054–29063.

Head NE, Yu H. 2004. Cross-sectional analysis of clinical and environmental isolates of Pseudomonas aeruginosa: biofilm formation, virulence, and genome diversity. *Infect Immun* 69:133–44.

Hentzer M, Wu H, Andersen JB, Riedel K, Rasmussen TB, Bagge N, Kumar N, Schembri MA, Song Z, Kristoffersen P, Manefield M, Costerton JW, Molin S, Eberl L, Steinberg P, Kjelleberg S, Hoiby N, Givskov M. 2003. Attenuation of Pseudomonas aeruginosa virulence by quorum sensing inhibitors. *The EMBO Journal* 22(15):3803-3815.

Hessle C, Andersson B, Wold A. 2000. Gram-positive bacteria are potent inducers of monocytic IL-12 while gram-negative bacteria preferentially stimulate IL-10 production. *Infect Immun* 68:3581–6.

Hessle C, Hanson LA, Wold AE. 1999. Lactobacilli from human gastrointestinal mucosa are strong stimulators of IL-12 production. *Clin Exp Immunol* 116:276-82.

Hessle CC, Andersson B, Wold AE. 2003. Gram-Negative, but Not Gram-Positive, Bacteria Elicit Strong PGE2 Production in Human Monocytes. Inflammation 27: 329-332.

Howell-Jones RS, Wilson MJ, Hill KE, Howard AJ, Price PE, Thomas DW. 2005. A review of the microbiology, antibiotic usage and resistance in chronic skin wounds. *J Antimicrob Chemother* 55: 143–149

Karlsson H, Larsson P, Wold AE, Rudin A. 2004. Pattern of cytokine responses to Gram-positive and gram-negative commensal bacteria is profoundly changed when monocytes differentiate into dendritic cells. *Infect Immun* 72:2671–2678.

Karlsson H, Hessle C, Rudin A. 2002. Innate Immune Responses of Human Neonatal Cells to Bacteria from the Normal Gastrointestinal Flora. *Infect Immun* 70: 6688–6696.

Kanzler H, Barrat FJ, Hessel EM, Coffman RL. 2007. Therapeutic targeting of innate immunity with Toll-like receptor agonists and antagonists. *Nature Medicine* 13: 552-559.

Kaper JB, Sperandio V. 2005. Bacterial cell-to-cell signaling in the gastrointestinal tract. *Infect. Immun* 73:3197–3209.

Kawai T, Akira S. 2005. Pathogen recognition with Toll-like receptors. *Curr Opin Immunol* 17:338–344.

Kievit TR, Iglewski BH. 2000. Bacterial quorum sensing in pathogenesis relationships. *Infect Immun* 68:4839–49.

Kleerebezem M, Boekhorst J, van Kranenburg R, Molenaar D, Kuipers OP, Leer R, Tarchini R, Peters SA, Sandbrink HM, Fiers MWEJ, Stiekema W, Lankhorst RMK, Bron PA, Hoffer SM, Groot MNN, Kerkhoven R, de Vries M, Ursing B, de Vos WM, Siezen RJ. 2003. Complete genome sequence of *Lactobacillus plantarum* WCFS1. *Proc. Natl. Acad. Sci. USA* 100:1990–1995.

Lam J, Chan R, Tam K, Costerton JW. 1980. Production of mucoid microcolonies by Pseudomonas aeruginosa within infected lungs in cystic fibrosis. *Infect. Immun.* 28: 546-556.

Landis SJ. 2008. Chronic Wound Infection and Antimicrobial Use. *Adv Skin Wound Care* 21:531–40.

Lardner A. 2001. The effect of extracellular pH on immune function. *J Leukoc Biol* 69:522–30.

Lazarus GS, Cooper DM, Knighton DR, et al. 1994. Definitions and guidelines for assessment of wounds and evaluation of healing. *Arch Dermatol* 130:489- 93.

Lebeer S, Claes IJJ, Verhoeven TLA, Shen C, Lambrichts I, Ceuppens JL, Vanderleyden J, De Keersmaecker SCJ. 2008. Impact of *luxS* and suppressor mutations on the gastrointestinal transit of *Lactobacillus rhamnosus* GG. *Appl. Environ. Microbiol.* 74:4711–4718.

Lebeer S, De Keersmaecker SCJ, Verhoeven TLA, Fadda AA, Marchal K, Vanderleyden J. 2007. Functional analysis of *luxS* in the probiotic strain *Lactobacillus rhamnosus* GG reveals a central metabolic role important for growth and biofilm formation. *J. Bacteriol.* 189:860–871.

Le Bouguenec C. 2005. Adhesins and invasins of pathogenic *Escherichia coli*. *Int. J. Med. Microbiol.* 295:471–478.

Lenz DH *et al.* 2004. The small RNA chaperone Hfq and multiple small RNAs control quorum sensing in *Vibrio harveyi* and *Vibrio cholerae*. *Cell* 118: 69–82.

Lewis K. 2001. Riddle of biofilm resistance. *Antimicrob Agents Chemother* 45:999–1007.

Llamas I; Quesada E, Martínez-Cánovas MJ, Gronquist M, Eberhard A, González JE. 2005. Quorum sensing in halophilic bacteria: detection of N-acyl-homoserine lactones in the exopolysaccharide-producing species of Halomonas. *Extremophiles* 9:333–341.

Mack DR, Ahrne S, Hyde L, Wei S, Hollingsworth MA. 2003. Extracellular MUC3 mucin secretion follows adherence of *Lactobacillus* strains to intestinal epithelial cells in vitro. *Gut* 52:827–833.

Madlener M, Parks WC, Werner S. 1998. Matrix metalloproteinases (MMPs) and their physiological inhibitors (TIMPs) are differentially expressed during excisional skin wound repair. *Exp Cell Res* 242:201-10.

Manefield M, Bovbjerg T, Rasmussen TB, Henzter M, Andersen JB, Steinberg P, Kjelleberg S, Givskov M. 2002. Halogenated furanones inhibit quorum sensing through accelerated LuxR turnover *Microbiology* 148:1119–1127.

Marco ML, Bongers RS, de Vos WM, Kleerebezem M. 2007. Spatial and temporal expression of *Lactobacillus plantarum* genes in the gastrointestinal tracts of mice. *Appl. Environ. Microbiol.* 73:124–132.

Matzinger P. 2007. Friendly and dangerous signals: is the tissue in control? *Nat Immunol* 8: 11-13.

Mayer-Scholl A, Averhoff P, Zychlinsk A. 2004. How do neutrophils and pathogens interact? *Current Opinion in Microbiology*, 7:62–66.

Menke NB, Ward KR, Witten TM, Bonchev DG, Diegelmann RF. 2007. Impaired wound healing. *Clin Dermatol* 25: 19–25.

Miller CH, Duerre JA. 1968. S-ribosylhomocysteine cleavage enzyme from *Escherichia coli. J. Biol. Chem.*243:92–97.

Miller JH. 1992. A short course in bacterial genetics. Cold Spring Harbor, NY: Cold Spring Harbor Laboratory Press.

Moon HS. 2004. Identification of Homoserine Lactone Derivatives Using the Methionine Functionalized Solid Phase Synthesis by Gas Chromatography/Mass Spectrometry. *Arch Pharm Res* 27 (1): 25-30.

Morohoshi T, Kato M, Fukamachi K, Kato N, Ikeda T. 2008. N-Acylhomoserine lactone regulatesviolacein production in Chromobacterium violaceum type strainATCC12472. *FEMS Microbiol Lett* 279:124–130.

Muller DM, Carrasco MS, Tonarelli GG, Simonetta AC. 2009. Characterization and purification of a new bacteriocin with a broad inhibitory spectrum produced by Lactobacillus plantarum lp 31 strain isolated from dry-fermented sausage *Journal of Applied Microbiology* 106: 2031–2040.

Muzio M, Bosisio D, Polentarutti N, D'amico G, Stoppacciaro A, Mancinelli R, van't Veer C, Penton-Rol G, Ruco LP, Allavena P, Mantovani A 2000. Differential expression and regulation of Toll-like receptors (TLR) in human leukocytes: selective expression of TLR3 in dendritic cells. *J Immunol* 164: 5998-6004.

Naumann D, Helm D, Labischinski H. 1991. Microbiological characterizations by FT-IR spectroscopy. *Nature* 351:81–82.

Nau R, Eiffert H. 2005. Minimizing the release of proinflammatory and toxic bacterial products within the host: A promising approach to improve outcome in life-threatening infections. *FEMS Immunol. and Medical Microbiol* 44:1–16.

Nedvidek W, Led F, Fischer P. 1992. Detection of 5-hydroxymethyl-1-2-methyl-3(2H)-furanone and of α-dicarbonyl compounds in reaction mixtures of hexoses and pentoses with different amines. *Z. Lebensm. Unters. Forsch.* 194: 222–228.

Nichols P, Henson J, Guckert J, Nivens DE, White DC. 1985. Fourier transform-infrared spectroscopic methods for microbial ecology: analysis of bacteria–polymer mixtures and biofilms. *J Microbiol Methods* 4:79–94.

Nickel JC, Ruseska I, Wright JB, Costerton JW. 1985. Tobramycin resistance of Pseudomonas aeruginosa cells growing as a biofilm on urinary catheter material. *Antimic. Agents Chemoth.* 27: 619-624.

Nivens DE, Ohman DE, Williams J, Franklin MJ. 2001. Role of Alginate and Its O Acetylation in Formation of *Pseudomonas aeruginosa* Microcolonies and Biofilms. *Journal of Bacteriology* 183 (3): 1047–1057.

Nivens DE, Palmer JR Jr, White DC. 1995. Continuous nondestructive monitoring of microbial biofilms: a review of analytical techniques. *J Ind Microbiol* 15:263–276.

Ochsner UA, Fietcher A, Reiser J. 1994. Isolation, characterization and expression in Escherichia coli of the Pseudomonas aeruginosa rhlAB genes encoding a rhamnosyltransferase involved in rhamnolipid biosurfactant synthesis. *J. Biol. Chem.* 269: 19787-19795.

Ochsner UA, Koch K, Fietcher A, J. Reiser J. 1994. Isolation and characterization of a regulatory gene affecting rhamnolipid biosurfactant synthesis in Pseudomonas aeruginosa. *J Bacteriol.* 176: 2044-2054.

O'Toole GA, Kolter R. 1998. The initiation of biofilm formation in Pseudomonas fluorescens WCS365 proceeds via multiple, convergent signaling pathways: a genetic analysis. *Mol Microbiol* 28:449–61.

Pearson JP, Pesci EC, Iglewski BH. 1997. Roles of Pseudomonas aeruginosa las and rhl quorum-sensing systems in control of elastase and rhamnolipid biosynthesis. *J Bacteriol.* 179: 5756-5767.

Peral MC, Rachid MM, Gobbato N, Huaman Martínez MA, Valdez JC. 2009. Interleukin-8 production by polymorphonuclear leukocytes from patients with chronic infected leg ulcers treated with Lactobacillus plantarum. *Clin Microbiol Infect* [Electronic publication ahead of print]

Peral MC, Huaman Martínez MA, Valdez JC. 2009. Bacteriotherapy with Lactobacillus plantarum in burns. *Int Wound J* 6:73–81.

Perdigon G, Valdez JC, Rachid M. 1998. Antitumour activity of yogurt: study of possible immune mechanisms. *J Dairy Res* 65:129-38.

Percival SL, Bowler P. 2004. Biofilms and their potential role in wound healing. *Wounds* 16: 234–240

Pesci EC, Iglewski BH. 1997. The chain of command in Pseudomonas quorum sensing. *Trends Microbiol.* 5: 132-135.

Pesci EC, Pearson JP, Seed PC, Iglewski BH. 1997. Regulation of las and rhl quorum sensing in Pseudomonas aeruginosa. *J Bacteriol.* 179: 3127-3132.

Prasad A, Cevallos ME, Riosa S, Darouiche RO, Trautner BW. 2009. A bacterial interference strategy for prevention of UTI in persons practicing intermittent catheterization. *Spinal Cord* 47:565-569.

Presser, KA, Ratkowsky DA, Ross T. 1997. Modelling the growth rate of *Escherichia coli* as a function of pH and lactic acid concentration. *Appl. Environ. Microbiol.* 63:2355–2360.

Pretzer G, Snel J, Molenaar D, Wiersma A, Bron PA, Lambert J, de Vos WM, van der Meer R, Smits MA, Kleerebezem M. 2005. Biodiversitybased identification and functional characterization of the mannose-specific adhesin of *Lactobacillus plantarum*. *J. Bacteriol.* 187:6128–6136.

Ramos AN, Peral MC, Valdez JC. 2008. Differences between Pseudomonas aeruginosa in a clinical sample and in a colony isolated from it: comparison of virulence capacity and susceptibility of biofilm to inhibitors. *Comp Immunol Microbiol Infect Dis* [Electronic publication ahead of print]

Ramos AN, Gobbato N, Rachid M, González L, Yantorno O, Valdez JC. 2010. Effect of Lactobacillus plantarum and Pseudomonas aeruginosa culture supernatants on polymorphonuclear damage and inflammatory response. Int Immunopharmacol. (2):247-51.

Reid G. 2001. Probiotic agents to protect the urogenital tract against infection. *Am J Clin Nutr* 73:437S–43S.

Reid, G, Kim SO, Kohler G. 2006. Selection, testing and understanding probiotic microbes. 46: 149-157. *FEMS Immunol Med Microbiol* 46: 149–157.

Reid G, Jass J, Sebulsky MT, McCormicK JK 2002. Potential uses of probiotic in clinical practice. *Clin Microbiol Rev.* 16:658–72.

Renelli M, Valério M, Lo RY, Beveridge TJ. 2004. DNA-containing membrane vesicles of *Pseudomonas aeruginosa* PAO1 and their genetic transformation potential. *Microbiology* 150:2161–93.

Roos K, Håkansson EG, Holm S. 2001. Effect of recolonisation with "interfering" a-streptococci on recurrences of acute and secretory otitis media in children; randomised placebo controlled trial. *BMJ* 322: 210-212.

Rumbaugh KP, Griswold JA, Iglewski BH, Hamood AN. 1999. Contribution of quorum sensing to the virulence of Pseudomonas aeruginosa in burn wound infections. *Infect Immun.* 67: 5854–5862.

Rusland M, Janeway Jr CA. 2002. Decoding the patterns of self and nonself by innate immune system. *Science* 296: 298-300.

Saier MH, Hancock REW, Lory S, Olson MV. 2000. Complete genome sequence of Pseudomonas aeruginosa PAO1, an opportunistic pathogen. *Nature*, 406: 959-964.

Salmond GPC, Bycroft BW, Stewart GSAB, Williams P. 1995. The bacterial 'enigma': cracking the code of cell-cell communication. *Mol Microbiol.* 16: 615-624.

Sandoz KM, Mitzimberg SM, and Schuster M. 2007. Social cheating in Pseudomonas aeruginosa quorum sensing. *Proc Natl Acad Sci USA.* 104: 15876–15881

Schmitt J, Flemming HC. 1998. FTIR-spectroscopy in microbial and material analysis. *Int Biodeterior Biodegrad* 41:1–11.

Servin AL. 2004. Antagonistic activities of lactobacilli and bifidobacteria against microbial pathogens. *FEMS Microbiol. Rev.* 28:405–440.

Sherjan CN, Savill J. 2005. Resolution of inflammmation: the beginning programs the end. *Nat Immunol* 12: 1191-1197.

Sigrid C, DeKeersmaecker J, Vanderleyden J. 2003. Constraints on detection of autoinducer-2 (AI-2) signalling molecules using Vibrio harveyi as a reporter. *Microbiology.* 149: 1953-1956.

Sloss JM, Cumberland N, Milner SM. 1993. Acetic acid used for the elimination of Pseudomonas aeruginosa from burn and soft tissue wounds. *J R Army Med Corps* 139:49–51.

Smith RS, Harris SG, Phipps R, Iglewski BH. 2002. The Pseudomonas aeruginosa quorum sensing molecules N-3-Oxododecanoyl homoserine lactone contributes to virulence and induces inflammation in vivo. *J Bacteriol.* 184: 1132–1139

Smith RS, Iglewski BH. 2003. Pseudomonas aeruginosa quórum sensing as potencial antimicrobial target. *J Clin Invest* 112: 1460-1465.

Stadelmann WK, Digenis AG, Tobin GR. 1998. Impediments to Wound Healing. *Am J Surg* 176 (Suppl 2A):39S–47S.

Starley IF, Mohammed P, Schneider G, Bickerler SW. 1999. The treatment of paediatric burns using topical papaya. *Burns* 25:636–9.

Sturme MHJ, Francke C, Siezen RJ, de Vos WM, Kleerebezem M. 2007. Making sense of quorum sensing in lactobacilli: a special focus on *Lactobacillus plantarum* WCFS1. *Microbiology* 153:3939–3947.

Sun J, Daniel R, Wagner-Döbler I, Zeng AP. 2004. Is autoinducer-2 a universal signal for interspecies communication: a comparative genomic and phylogenetic analysis of the synthesis and signal transduction pathways. *BMC Evolutionary Biology* 4:36.

Taga ME, Miller ST, Bassler BL. 2003. Lsr-mediated transport and processing of AI-2 in *Salmonella typhimurium*. *Mol. Microbiol.* 50: 1411–1427.

Tagg JR, Dierksen KP. 2003. Bacterial replacement therapy: adapting 'germ warfare' to infection prevention. *Trends in Biotechnol* 21: 217-223.

Tateda K, Ishii Y, Horikawa M, Matsumoto T, Miyairi S, Pechere JC, et al. 2003. The Pseudomonas aeruginosa autoinducer N-3-oxododecanoyl homoserine lactone accelerates apoptosis in macrophages and neutrophils. *Infect Immun* 71:5785–93.

Trevani AS, Andonegui G, Giordano M, López DH, Gamberale R, Minucci F, et al. 1999. Extracellular acidification induces human neutrophil activation. *J Immunol* 162:4849–57.

Turovskiy Y, Chikindas ML. 2006. Autoinducer-2 bioassay is a qualitative, not quantitative method influenced by glucose. *Journal of Microbiological Methods* 66:497–503.

Usher LR, Lawson RA, Geary I, Taylor CJ, Bingle CD, Taylor GW, et al. 2002. Induction of neutrophil apoptosis by the Pseudomonas aeruginosa exotoxin pyocyanin: a potential mechanism of persistent infection. *J Immunol* 168:1861–8.

Valdez JC, Peral MC, Rachid M, Santana M, Perdigon G. 2005. Interference of Lactobacillus plantarum on Pseudomonas aeruginosa in vitro and in infected burns. The potential use of probiotic in wound treatment. *Clin Microbiol Infect* 11:472–9.

Valdéz JC, Rachid M, Gobbato N, Perdigón G. 2001. Lactic acid bacteria induce apoptosis inhibition in Salmonella typhimurium infected macrophages. *Food Agricult Immunol* 13: 189-197.

Vermes I, Haanen C, Reutelingsperger CPM. 1995. A novel assay for apoptosis: flow cytometric detection of phosphatidylserine expression on early apoptotic cells using fluorescein labeled Annexin V. *J Immunol Methods* 180:39–52.

Vendeville A, Winzer K, Heurlier K, Tang CM, Hardie KR. 2005. Making 'sense' of metabolism: Autoinducer-2, luxS and pathogenic bacteria. *Nat Rev Microbiol* 3: 383-396.

van Baarlena P, Troosta FJ, van Hemerta S, van der Meera C, de Vose WM, de Groota PJ, Hooivelda GJEJ, Brummera RJM, and Kleerebezema M. 2009. Differential NFkB pathways induction by Lactobacillus plantarum in the duodenum of healthy humans correlating with immune tolerance. *Proc Natl Acad Sci USA*. 106: 2371–2376.

von Bodman SB, Willey JM, Diggle SP. 2008. Cell-Cell Communication in acteria: *United We Stand. J Bacteriol* 190: 4377–4391.

Webb JS, Thompson LS, James S, Charlton T, Tolker-Nielsen T, Koch B, et al. 2003. Cell death in Pseudomonas aeruginosa biofilm development. *J Bacteriol* 185:4585–92.

Whiteley M, Lee KM, Greenberg EP. 1999. Identification of genes controlled by quorum sensing in Pseudomonas aeruginosa. *PNAS* 96(24): 13904–13909.

Winzer K, Hardie KR, Williams P. 2003. LuxS and autoinducer-2: their contribution to quorum sensing and metabolism in bacteria. *Adv. Appl. Microbiol.* 53: 291–396.

Zaborina O, Dhiman N, Chen ML, Kostal J, Holder IA, Chakrabarty AM. 2000. Secreted products of a nonmucoid Pseudomonas aeruginosa strain induce two modes of macrophage killing: external-ATP-dependent, P2Z-receptor-mediated necrosis and ATP-independent, caspase-mediated apoptosis. *Microbiology* 146:2521–30.

Zhu J, Chai Y, Zhong Z, Li S, Winans SC. 2003. Agrobacterium Bioassay Strain for Ultrasensitive Detection of *N*-Acylhomoserine Lactone-Type Quorum-Sensing Molecules: Detection of Autoinducers in *Mesorhizobium huakuii*. *Aplied and environmental microbiology*. Nov: 6949–6953.

INDEX

A

acetic acid, 44, 46, 47, 51, 74
acid, 1, 8, 15, 19, 23, 26, 35, 38, 39, 40, 44, 47, 48, 51, 79, 89, 90
acidity, viii, 16, 17, 19, 24, 25, 26, 35, 41, 49, 51
acidosis, 46
adhesion, 18, 20, 21, 25, 41, 73
agar, 24, 31, 53, 54, 56
AIDS, 13
aldolase, 37
algae, 4
amines, 86
amino acids, 4, 45
angiogenesis, 52
antibiotic, 3, 43, 59, 84
antibiotic resistance, 43, 59
antibody, 10
antigen, 6, 10, 11
antigen-presenting cell, 6
anti-inflammatory medications, 2
antimicrobial therapy, 59
apoptosis, 1, 6, 10, 13, 44, 46, 48, 49, 50, 51, 52, 56, 57, 65, 67, 72, 74, 75, 90, 91
apoptotic mechanisms, 11
architecture, 2
Argentina, 11, 44, 62, 66, 77
assessment, 85
ATP, 42, 91

attachment, vii, 40
authors, vii, 11, 51, 52
autolysis, 45

B

bacterial infection, 52, 72, 81
bacterial strains, 7, 8
bacteriocins, 39, 82
bacterium, 31
barriers, 59
behaviors, 4
bioassay, 32, 33, 35, 41, 90
biological responses, 9
bioluminescence, 35, 37, 41
biomass, 18, 20, 21, 22
biopsy, 11, 53, 66
biosurfactant, 28, 87
biosynthesis, 87
biotechnology, 4
bleeding, 66
blood clot, 1
blood vessels, 2
boric acid, 34, 35
breakdown, 64
burn, 53, 54, 55, 56, 57, 58, 59, 60, 61, 62, 63, 79, 82, 88, 89
burn wound surface, 60
by-products, vii

C

cancer, 13
carbohydrates, 45
carcinogen, 10
caregivers, 2
catheter, 8, 87
cecum, 40, 42
cell culture, 32, 38
cell death, 6
cell signaling, 41, 84
cell surface, 5
Census, 82
chain of command, 87
chemokines, 1, 5
chromatograms, 31
chromatography, 31, 40
City, 11
cleavage, 86
clone, 53
closure, 2, 67
coding, 4
collagen, 2, 64
colon, 10, 40
colonisation, 7, 63
colonization, 3
color, iv, 45
communication, viii, 4, 33, 41, 72, 73, 82, 89
community, 41
competition, 28, 40, 41, 42
competitive conditions, 42
complement, 5
complications, 72
composition, 18, 23, 28
compounds, 5, 27, 32, 39, 86
connective tissue, 50, 56, 64
consensus, 82
consent, 66
consumption, 11
contamination, 2
control group, 7, 50, 54
copyright, iv
correlation, 46
cost, vii, 2, 59, 60, 64, 75
culture, 15, 16, 18, 20, 28, 30, 31, 32, 36, 38, 41, 51, 53, 54, 62, 65, 66, 74, 88
cystic fibrosis, 13, 72, 81, 83, 84
cytokines, 5, 9, 10, 11, 45, 52, 65, 83
cytoplasm, 29, 36, 41
cytotoxic agents, 1
cytotoxicity, 9, 51, 52, 53

D

damages, iv
danger, 6, 10
debridement, 3, 59, 65, 71, 74
defects, 42
defence, 6, 64
defense mechanisms, 8, 9
deficiencies, 59
degradation, 64
dendritic cell, 6, 84, 86
Denmark, 45
deposition, 2
dermis, 50
detection, 4, 5, 85, 89, 90
developed countries, 64
diabetes, 2, 43, 64
diabetic patients, 65, 66, 67, 69, 70, 71, 74
diet, 2
digestion, 56
discomfort, 66
discrimination, 5
displacement, 40
distilled water, 22
diversity, 41, 83
DNA, 4, 5, 16, 17, 19, 26, 34, 56, 88
donors, 66
dosage, 3
dressings, 3, 60, 65, 82
drug design, 7
duodenum, 90

E

ecology, 2, 83, 86
edema, 50

Index

elastin, 37
ELISA, 45, 81
emission, 46, 48, 79
encoding, 11, 39, 87
endothelial cells, 2, 52, 80, 83
environmental conditions, 26
enzymes, 1, 4, 26, 28, 29, 32, 34, 37, 43, 52, 64
epidermis, 2, 50
epithelia, 6
epithelial cells, 2, 41, 85
epithelium, 80
excision, 59, 61
excitation, 46, 48, 79
exclusion, 40
exercise, 8
exopolysaccharides, 16
exposure, 1, 3, 6
extracellular matrix, 2, 3, 15, 64
extraction, 30, 40
exudate, 62

F

fatty acids, 82
fears, 7
fever, 45
fibers, 2
fibrinogen, 6
fibroblast growth factor, 52
fibroblasts, 2, 43, 51, 64, 65, 72, 75, 80
fibrosis, 2
filters, 44
filtration, 40, 44
flagellum, 13
flora, 7
fluorescence, 46, 56, 79
formaldehyde, 17
FTIR, 89
fungi, 5, 60

G

gastrointestinal tract, 40, 80, 84, 85

gel, 40
gene expression, 27, 41, 81, 82
gene promoter, 4
genes, 3, 4, 11, 26, 27, 28, 40, 80, 82, 85, 87, 91
genetics, 86
genome, 28, 83, 84, 89
Germany, 46
glucose, 35, 90
glycerol, 53
GPC, 89
granules, 1, 17, 64
growth factor, 1, 52, 65, 81, 83
growth rate, 88
guidelines, 85

H

habitats, 72
heat shock protein, 6
height, 5
homocysteine, 34, 42
host, vii, 3, 4, 5, 8, 16, 40, 43, 58, 60, 64, 72, 82, 86
hydrogen, 80
hydrogen peroxide, 80
hydrolysis, 34
hydroxyl, 80
hypothesis, 26, 40, 42, 80
hypoxia, 2

I

IFN, 9
IL-13, 5
IL-8, viii, 5, 51, 65, 66, 67, 69, 72, 74, 75, 79
immune function, 85
immune response, 4, 5, 10, 11, 60, 62, 64
immune system, 3, 5, 8, 9, 10, 45, 52, 72, 80, 88
immunity, 39
immunocompetent cells, 8
immunocompromised, 13

immunoglobulin, 45
immunomodulation, 9, 10
immunomodulatory, 8, 10, 11
immunostimulatory, 83
in vivo, vii, 9, 10, 11, 13, 17, 42, 50, 52, 63, 72, 74, 89
incidence, 60
incubation time, 49, 51, 74
induction, 35, 52, 81, 90
inflammation, 1, 6, 52, 53, 64, 65, 72, 80, 89
inflammatory cells, 62
inflammatory mediators, 43, 64
inflammatory responses, 6, 10
ingestion, 11
inhibition, 8, 10, 11, 15, 16, 19, 23, 25, 26, 28, 29, 30, 32, 35, 39, 72, 73, 90
inhibitor, 16, 29, 30
initiation, 87
injections, 10
innate immunity, 5, 84
inoculation, 8, 13, 50, 74
interface, 40
interference, vii, viii, 6, 7, 8, 13, 62, 67, 72, 73, 81, 88
interferon, 9
interleukin-8, 1
intestinal tract, 80
iodine, 60
IR spectra, 22, 23, 26
IR spectroscopy, 86
iron, 39, 81
ischemia, 43
isolation, 53

J

jejunum, 42
joints, 4

K

keratinocytes, 43, 51, 80, 83
kinetic curves, 20

kinetics, 19, 20, 21, 22, 23, 26

L

lactic acid, viii, 8, 19, 20, 24, 25, 26, 39, 51, 82, 88
lactobacillus, 10
latency, 21, 25, 73
leakage, 39
lesions, 66
lifetime, 64
ligand, 34
limitations, 65
lipids, 1
lipoproteins, 9
Listeria monocytogenes, 40
liver, 55, 56
localization, 60
locus, 40
lymphoid, 8
lymphoid tissue, 8
lysis, 5

M

macromolecules, 64
macrophages, 1, 5, 9, 10, 14, 45, 52, 64, 72, 75, 81, 90
majority, 3
malignancy, 66
malnutrition, 43
management, 59
mass spectrometry, 31, 40
matrix, vii, 3, 5, 15, 16, 18, 19, 23, 26, 29, 64
matrix metalloproteinase, 5, 64
media, 7, 16, 67
median, 67, 70, 71
medicines, 60
membranes, 5
metabolic disorder, 2
metabolism, 3, 16, 42, 90, 91
metabolites, 28, 34, 39
methanol, 31

methylation, 34
mice, vii, 10, 13, 40, 50, 51, 52, 53, 54, 55, 56, 80, 85
microbial cells, 3, 19
microorganism, 15, 56
microscope, 56
microscopy, 50
MMP, 64
MMPs, 64, 85
modification, viii, 71
moisture, 65
molecular structure, 10
molecular weight, 40
molecules, 4, 5, 8, 10, 21, 29, 30, 31, 34, 39, 41, 51, 73, 83, 89
monitoring, 19, 87
Moon, 31, 86
morphology, 56
mRNA, 37
mucin, 85
mucoid, 19, 22, 84
mucosa, 11, 84
muscular tissue, 56
mutant, 28, 32, 40, 41, 42, 44

N

NaCl, 19, 23, 24, 45
necrosis, 6, 13, 44, 46, 49, 50, 51, 58, 65, 72, 74, 75, 91
neonates, 59
neurogenic bladder, 79
neutrophils, 1, 5, 14, 43, 47, 48, 49, 64, 72, 74, 81, 86, 90
nitric oxide, 5
nitrogen, 28
NK cells, 9
normal curve, 24
nuclei, 56
nutrients, 3, 28, 41
nutritional deficiencies, 2

O

oligomerization, 10
oligosaccharide, 40
omega-3, 1
operon, 36
organ, 2, 6, 8
organism, 4, 34
otitis media, 7, 88
overproduction, 40
oxygen, 1, 15, 43, 51, 52

P

pathogenesis, 3, 4, 27, 84
pathogens, vii, 4, 6, 8, 10, 11, 12, 40, 43, 60, 62, 72, 86, 89
pathophysiology, 64
pathways, 9, 11, 89, 90
pattern recognition, 5, 9
PBMC, 8, 9
peptides, 1, 45
performance, 42, 53
peripheral blood, 9, 43, 65, 66, 69, 71
peripheral blood mononuclear cell, 9
permeability, 81
permission, iv
phagocytosis, 4, 51, 56, 57, 72, 74, 75
phenotype, 42
phosphates, 37
phosphatidylserine, 90
phosphorylation, 4
physiology, 5
placebo, 88
plasma cells, 10
plasmolysis, 45
plastic surgeon, 66
platelets, 1
polymer, 86
polymers, vii
polypropylene, 19, 20, 22, 23
polyvinyl chloride, 62, 69
population density, 4, 39, 41
potassium, 46

prevention, 60, 81, 88, 90
probiotic, 7, 8, 9, 40, 42, 79, 81, 85, 88, 90
producers, 60, 64, 69
pro-inflammatory, 45
proliferation, 65
promoter, 4
prophylaxis, 59, 63
prostaglandins, 1
proteases, 26, 27, 29, 30
proteinase, 64
proteins, 4, 16, 26, 34, 36
proteolytic enzyme, 40
Pseudomonas aeruginosa, 8, 13, 26, 38, 57, 58, 60, 71, 79, 80, 81, 82, 83, 84, 87, 88, 89, 90, 91
purification, 86

Q

quality of life, vii, 2

R

radiation, 19
radicals, 43, 52, 80
reactions, 34, 37
reactive oxygen, 1, 4
receptor sites, 40
receptors, 5, 9, 36, 42, 73, 81, 84, 86
recognition, 5, 10, 84
recombinant DNA, 16
recommendations, iv
recruiting, 64
regeneration, 58
relapses, 3
relatives, 10
relief, 65
repair, viii, 52, 53, 58, 62, 72, 75, 85
replacement, 90
repression, 37
repressor, 36
reproductive age, 7
residues, 40
resistance, 3, 42, 43, 60, 84, 85, 87

resolution, 1
respect, 20, 60, 72
rhamnolipid, 28, 39, 58, 87
rights, iv
RNA, 6, 34, 85
room temperature, 48

S

saturation, 18, 21
scatter, 46
secrete, 9
secretion, 9, 14, 16, 45, 64, 81, 85
sedimentation, 46
seeding, 24
sensation, 66
sensing, vii, viii, 3, 4, 7, 8, 13, 15, 27, 28, 39, 43, 51, 53, 58, 72, 73, 80, 82, 83, 84, 85, 87, 88, 89, 91
sensitivity, 16
serine, 64
serum, 66
side effects, 7
signal transduction, 4, 5, 89
signaling pathway, 42, 87
signalling, 9, 37, 89
signals, viii, 3, 4, 6, 10, 13, 39, 41, 53, 72, 73, 81, 85
signs, 2, 4, 66
silver, 79
skin, 7, 13, 50, 51, 52, 54, 55, 56, 60, 74, 81, 84, 85
small intestine, 40
species, viii, 1, 2, 4, 9, 41, 69, 83, 85
spectroscopy, 18, 89
spinal cord, 8, 79
spinal cord injury, 8
spleen, 55, 56
Spring, 86
stasis, 64
sterile, 2, 8, 54, 60, 64
stomach, 42
strategy, vii, 6, 42, 72, 88
streptococci, 7, 60, 88
stretching, 19

structural characteristics, 10
substrates, 15, 37
Sun, 32, 89
suppression, 60
surface area, 60
surgical debridement, 59, 60
survival, 3, 4, 54
susceptibility, 88
suspensions, 23, 46, 48, 56
synthesis, 3, 4, 28, 32, 33, 53, 73, 80, 87, 89

T

T cell, 9, 11
temperature, 54
tension, 51
testing, 88
therapy, 8, 60, 90
thorax, 61
tissue, viii, 1, 2, 50, 51, 52, 53, 54, 56, 58, 60, 61, 62, 63, 64, 65, 66, 67, 69, 72, 74, 75, 85, 89
TLR, 5, 10, 86
TLR2, 5, 9
TLR3, 86
TLR4, 5, 9, 45
TLR9, 5
TNF, 9, 10, 11, 79
TNF-α, 9, 10, 11
tonsillitis, 7
toxin, 58
transcription, 4, 6, 9, 27, 28, 37
transcription factors, 9
transduction, 37
transferrin, 81
transformation, 21, 88
transforming growth factor, 65
transport, 4, 90
trauma, 1
trial, 62, 81, 88
tumor, 9
tumor necrosis factor, 9
tumour growth, 10
tumours, 10
turnover, 4, 85
type 2 diabetes, 65, 71

U

ulcer, vii, viii, 2, 17, 65, 66, 70, 71, 72, 74, 75
ultrasound, 66
uniform, 31
urinary tract, 79, 81
urinary tract infection, 79, 81

V

vacuum, 22
vagina, vii, 7, 8, 40, 72
variations, 19, 69
venipuncture, 45
vessels, 2
vibration, 19
viruses, 5
vitamins, 2, 45

W

water vapor, 22
wavelengths, 46
wells, 16, 29
white blood cell count, 50
wild type, 41
wound healing, vii, viii, 1, 2, 43, 51, 52, 59, 65, 86, 87
wound infection, 2, 54, 59, 60, 65, 72, 79, 88

Y

yeast, 45